Water Science and Technology

WATER SCIENCE and TECHNOLOGY

T. H. Y. Tebbutt
BSc, SM, MICE, MIWE, MASCE,
FIPHE, MInstWPC
Department of Civil Engineering,
University of Birmingham

BOOKS
10 East 53d St., New York 10022
(a division of Harper & Row Publishers, Inc.)

Published in the United States of America 1973
by Harper & Row Publishers, Inc.
Barnes & Noble Import Division

© T. H. Y. Tebbutt 1973

Printed in Great Britain

ISBN 06–496790–5

Preface

Recent concern about the state of our environment has aroused public interest in the subjects of water pollution and the development of water resources. In any meaningful discussion of such matters a basic understanding of the relevant science and technology is essential. This book aims to give a brief account of current practice in water supply and treatment and wastewater collection and treatment. It is intended as an introduction to the subject for the layman. Students and specialists in other fields who wish to gain a basic understanding of the problems to be found in water science and technology may also find it useful. In the hope that the book will stimulate further interest I have included a list of sources of more detailed information.

In preparing a book of this nature the work of many other authors must inevitably have influenced me and I trust that the reader will appreciate the debt which I owe to others.

Birmingham, 1972   T. H. Y. T.

Contents

1. Introduction — 1
2. Nature and Occurrence of Water — 4
3. Water Use — 17
4. Collection of Water — 29
5. Water Treatment — 53
6. Distribution of Water — 85
7. Collection of Wastewater — 99
8. Water Pollution — 109
9. Wastewater Treatment — 127
10. Water Reclamation — 157
11. Economic and Legal Aspects — 171
12. Research and Future Developments — 179

Conversions to Imperial units — 188

Further reading — 189

Index — 191

Plates

1.	Photograph of stained bacteria magnified about 800 times	*p.* 12
2.	Colonies of bacteria growing on agar medium	13
3.	MacConkey Broth tubes for counting coliform bacteria	15
4.	Propeller-type current meter	33
5.	Gauging weir	36
6.	Claerwen dam	41
7.	One of the Elan Valley dams	41
8.	Clywedog dam	43
9.	Balderhead dam	45
10.	An electrical analogue for groundwater flow	51
11.	Continuous band screen	56
12.	Microstrainers	56
13.	Jar-test apparatus	60
14.	Vertical flow sedimentation tank	64
15.	Circular flocculation and sedimentation tank	64
16.	Rapid sand filter	67
17.	Backwashing a rapid sand filter	68
18.	Air scour prior to backwashing	68
19.	Pressure filter	70
20.	Chlorinator installation	76
21.	Vertical spindle centrifugal pump	92
22.	Submersible pump	93
23.	A conventional reinforced-concrete water tower	98
24.	Morwick water tower	98
25.	Construction of a large sewer	103
26.	Sewage fungus in a polluted stream	113
27.	Bar screen	128
28.	Comminutor	129
29.	Control flume, grit channels and screens	131
30.	Stormwater overflow	131
31.	Stormwater settling tank	132

32.	Primary sedimentation tank	p. 132
33.	Circular bacteria bed and dosing chamber	135
34.	Rope-hauled distributor on rectangular bacteria bed	135
35.	Circular humus tank	136
36.	Diffused air activated sludge	139
37.	Diffused air tank under construction	139
38.	Cone aerator	140
39.	Brush aerator	140
40.	Circular final settling tank	141
41.	Interior arrangement of final settling tank	143
42.	Effluent as discharged from final settling tank	143
43.	Sludge drying beds at a small works	145
44.	Mechanical sludge lifting equipment	147
45.	Sludge digestion tank	147
46.	Filter presses	149
47.	Vacuum filter	149
48.	Packaged extended-aeration plant	152
49.	Pasveer ditch	155
50.	Flash distillation plant	165

PHOTOGRAPHIC ACKNOWLEDGMENTS

The author is most grateful to the following organizations for permission to reproduce the photographs noted:

Amalgamated Power Engineering Ltd, Plate 21
F. W. Brackett and Co. Ltd, Plate 11
Department of Civil Engineering, University of Birmingham, Plates 1, 2, 3, 10 and 25
Dorr-Oliver Co. Ltd, Plate 47
Edwards and Jones Ltd, Plate 46
Glenfield and Kennedy Ltd, Plate 12
Grundfos Pumps Ltd, Plate 22
Newcastle and Gateshead Water Company, Plate 24
Satec Ltd, Plate 48
Wallace and Tiernan Ltd, Plate 20
J. G. Weir Ltd, Plate 50

The other plates are from photographs by the author who deeply appreciates the way in which numerous authorities and undertakings have allowed him to visit and photograph their installations.

Figures in Text

1.	Aerobic oxidation	*p.*	8
2.	Anaerobic oxidation		8
3.	The biological cycle in surfacewater		10
4.	The hydrological cycle		15
5.	Types of cooling systems		24
6.	The standard rain gauge		30
7.	One type of recording rain gauge		31
8.	A typical rainfall intensity–duration curve		31
9.	Velocity distributions in a stream channel		32
10.	Current metering		34
11.	Typical analytical results from a chemical dilution gauging		35
12.	Streamflow hydrograph		36
13.	Typical hydrographs		37
14.	Effect of storage on yield		40
15.	A direct supply reservoir		41
16.	A river regulation reservoir		42
17.	The Clywedog regulation scheme for the river Severn		44
18.	Stratification in a lake		46
19.	Groundwater abstraction		49
20.	Intrusion of sea water into an aquifer due to overpumping near the coast		49
21.	Cone of depression round a well		50
22.	Induced re-charge of an aquifer from a river due to the effects of pumping on the water table		51
23.	Particle sizes and effective ranges of treatment processes		54
24.	Flow diagram for a conventional water treatment plant		55
25.	Rapid mixing		58
26.	Methods of flocculation		59
27.	Typical jar-test results		60
28.	Settling behaviour of discrete and flocculent particles		62
29.	Flow-through curves for sedimentation tanks		62

30.	Types of sedimentation tank	p. 63
31.	Slow sand filter	66
32.	Rapid gravity sand filter	67
33.	Rapid pressure filter application	69
34.	Filter head-loss curve	71
35.	Modified forms of filter	71
36.	Head loss characteristics of a mixed media bed	73
37.	The chlorine breakpoint curve	75
38.	Flow diagram for ozone disinfection	76
39.	Lime softening	79
40.	Elements of a water distribution system	85
41.	Pipe joints for water mains	88
42.	Optimum size of a pumping main	89
43.	Positioning of air and scour valves on a pipeline	90
44.	Diagrammatic arrangement of an air valve	91
45.	Typical pumping station	93
46.	Air-loaded vessel for surge prevention	94
47.	Surface feeder tanks for surge prevention	94
48.	Cathodic protection	96
49.	Function of a service reservoir	96
50.	Storm water overflow	100
51.	Special sewer sections	101
52.	Joints in sewer pipes	103
53.	Basic elements of a sewerage system	104
54.	A surfacewater sewerage system and its discharge hydrograph	106
55.	A time–area diagram for a surfacewater sewerage system	107
56.	The dissolved oxygen sag curve	115
57.	A sea outfall	121
58.	A bacteria bed	134
59.	The activated-sludge process	138
60.	A pebble-bed clarifier	142
61.	The importance of sludge dewatering	145
62.	Flow diagram for a conventional sewage treatment plant	151
63.	Re-use of water in a large river system	158
64.	Multi-stage flash distillation	165
65.	A solar still	166

66.	Principle of electrodialysis	p. 167
67.	Principle of reverse osmosis	169
68.	Flow diagram for vacuum freezing	170
69.	Catchment of the river Trent	180

Tables

1.	Types of micro-organisms	p.	11
2.	Typical characteristics of water from various sources		14
3.	Domestic water consumption in the UK		18
4.	World Health Organization drinking water standards		22
5.	Typical industrial water demands		25
6.	Typical quality requirements for industrial process water		26
7.	Typical wastewater characteristics		28
8.	Bilham rainfall intensities		32
9.	Probable treatment for water from various sources		54
10.	Effect of particle size on settling velocity		57
11.	Typical pipe roughness factors		87
12.	Typical impermeability factors		105
13.	Typical effluent standards for discharges to rivers		125
14.	Typical effluent standards for discharges to sewers		125
15.	Composition of sewage sludges compared with a fertilizer		150
16.	Treatment for various industrial wastewaters		155
17.	Quality degradation in one cycle, water to sewage effluent		158

1 *Introduction*

Water, which is almost the only compound to occur naturally on the earth's surface as a liquid has, until recently, been taken for granted with little thought given to its conservation. The concept of water as a natural resource is relatively new but nevertheless very necessary as growing populations demand ever increasing supplies of water to satisfy their requirements. The presence or absence of water in an area has a profound influence on its development and prosperity; over the centuries many skirmishes have taken place over supplies of water and water rights.

There is, of course, a vast amount of water in the world but much of it is unsuitable for consumption because of its high salinity. If the remainder were distributed in accordance with population density there would be ample water for everyone. Problems of water shortage arise because of the uneven distribution of water and population on the earth's surface and the polluting effects of waterborne wastes which have come to be an accepted part of modern civilization. Thus, in the majority of the most densely populated parts of the world, which often have lower than average rainfall, existing sources of fresh water are already under severe strain and will be unable to meet all demands made upon them in the near future. These impending shortages have focused attention on the general field of water resources and their conservation. The traditional concepts of water supply from single-purpose impounding reservoirs are being abandoned in favour of multi-purpose schemes which are designed to permit as many uses as possible of a water resource before it reaches the sea. Such multi-purpose use brings as a penalty the need to accept more polluted supplies as the source for drinking water so that more complex, and expensive, treatment methods become necessary. In addition, multi-purpose development has to resolve the conflicting requirements of different uses of water. A reservoir needs to be kept as full as possible to satisfy water supply demands, but should be kept empty to provide flood control—an obvious clash of interests!

The sea itself can no longer be dismissed as a possible source of drinking water and in arid parts of the world desalination techniques are increasingly being used for the production of potable supplies from saline waters.

In the days when demands for water were relatively small, the need for measurement of the quantity and quality was limited but as more and more of the available water is utilized an accurate assessment of our water resources is essential. The rapidly developing science of hydrology is specifically concerned with means of assessing the quantity of water resources but it should be realized that quantity and quality considerations are inseparable since one is meaningless without the other.

Research in the field of water technology is being carried out at an increasing pace by government and industrial research establishments, universities and private firms. By such research, covering both quality and quantity aspects of the subject, it is hoped that a more enlightened approach to water conservation will develop and that the ensuing technological advances will assure adequate supplies of water for all reasonable demands. One thing is certain however, water will become much more expensive because of the cost of developing new sources and the need to expand treatment processes and pollution prevention measures. More and more waste products are being discharged to water sources and their identification and removal from water supplies is likely to be a difficult and expensive procedure. Increasing costs for water may in themselves cause some reduction in the rate of increase of water demand but this effect is likely to be relatively small in the long term.

Valuable as water is to man's wellbeing, it should not be forgotten that drinking water can provide a vehicle for the spread of disease which can only be countered by constant vigilance on the part of engineers, scientists and doctors. It is a sobering fact that, despite the rarity of waterborne diseases in western countries, they are still rampant in other parts of the world. Five million babies die every year as the result of waterborne disease and one in six of the world's population suffer annually from such diseases.

Introduction

Strangely enough, although ancient civilizations in the Middle and Far East, and later the Romans, appreciated the value of wholesome water and hygienic waste disposal systems, little was done in this field in Europe for centuries.

The explosive growth of towns and cities as a result of the industrial revolution caused widespread pollution of water supplies and outbreaks of cholera and typhoid of increasing severity ensued. In the 1850s the Thames became so polluted that it was a public nuisance and it is recorded that Parliament had sometimes had occasion to adjourn its business because of the stench from the river. Since much of London's water supply was at that time obtained from the Thames or from shallow wells contaminated by leakage from old sewers there were many outbreaks of waterborne disease in the city. In 1854 an outbreak of cholera in London caused 10 000 deaths amongst the population of $2\frac{1}{2}$ million, and as a result the connection between polluted drinking water and cholera was established, although it was many years before the causative agent, a bacterium called *Vibrio cholerae*, could be isolated. Following this discovery of circumstantial evidence, measures were instituted to protect the quality of drinking water and the disease was eventually brought under control and finally eradicated from the British Isles.

Continued improvement in the science and technology of pollution prevention and water treatment is thus reflected in the absence of waterborne disease in the UK and most other developed countries. Every year, however, the growing demands on water and the inevitable increases in discharges of sewage and industrial wastes pose new problems which must be solved to ensure the continued supply of a safe and wholesome water. New methods of analysis and treatment have to be developed to keep abreast with new and exotic wastes produced by what has been called 'the effluent society' and it is clear that in the future, although we may not have to drink our own sewage, most of us will certainly be drinking other people's sewage. Water science and technology will ensure, however, that the original source is not apparent to the consumer and, more important, that the consumer suffers no ill effects due to the previous history of his drinking water.

2 Nature and Occurrence of Water

Water is such a widespread material that its presence is accepted without question and its importance is only really appreciated when there is a shortage. The chemical formula for water, H_2O, is probably the only formula which is almost universally known; in fact, with the exception of specially treated demineralized water, all samples contain varying amounts of other materials. Even the formula H_2O is not really true, as water normally exists as H_6O_3 or H_8O_4, particularly at low temperatures, and it is these more complicated structures which are responsible for the unusual properties of water. Thus, water has the unusual and often troublesome property of expanding on freezing due to a rearrangement of the molecular structure. On thawing, contraction occurs until a temperature of 3·94°C is reached, at which point water has its maximum density.

The other materials present in water are measured in terms of milligrammes per litre (mg/l) which is essentially equivalent to parts per million (ppm). The impurities may vary from a few mg/l in rain water to about 35 000 mg/l in seawater. Most river waters contain 100 to 300 mg/l of impurities and raw sewage contains about 1000 mg/l of impurities. It is thus important to appreciate that no natural water is chemically pure and, indeed, pure H_2O is not a palatable drink. It is the small concentrations of dissolved solids and gases which give most natural waters their pleasant tastes. On the other hand concentrations of less than 0·01 mg/l of phenol can render a water quite unpalatable, as happens when chlorophenols are formed following disinfection of the water with chlorine. Many other organic compounds, some natural but most man-made, can produce similar undesirable effects at low concentrations.

Cleanliness of water can thus only be a relative term and to obtain a true picture of the nature of a water it is necessary to make a much more detailed examination than simply deter-

Nature and Occurrence of Water

mining the total weight of impurities present. It is useful to specify the properties of a water sample with reference to its physical, chemical and biological characteristics.

PHYSICAL CHARACTERISTICS

The total solids present in a sample can be divided into dissolved and suspended forms—dissolved solids do not impair the clarity of a water although they often change its colour, whereas suspended solids give the water a cloudy or turbid appearance. A further division may be made by drying and weighing the solids. They are fired in a furnace at 600°C, which volatilizes the organic solids, leaving only a residue of inorganic solids to be weighed. The weight of organic solids is then calculated from the loss in weight on firing. The proportion of organic matter in a sample gives a useful indication of its origin: groundwater and a sewage effluent might both have total solids contents of, say, 700 mg/l, but the groundwater would have almost no organic solids whereas the effluent would contain about two thirds organic solids.

For domestic water supplies the appearance and taste of a sample are readily discernible to the consumer, and in potable water treatment care is taken to keep such properties within desirable limits, which are usually based, at least in part, on aesthetic considerations. Thus, a water with a high colour from an upland source is in fact probably quite safe to drink, but may give the consumer the impression that it is dangerous. He might then take water from a lowland stream which would have less colour, but which would be much more likely to contain harmful (but invisible) pollution in the form of dissolved chemicals or micro-organisms. Important physical characteristics are thus:

(a) *Colour.* Due to dissolved matter, usually organic acids similar to those which give tea its characteristic colour and which give natural upland waters a golden yellow colour.
(b) *Turbidity.* Presence of colloidal solids which may be clay or silt particles or micro-organisms.
(c) *Taste and odour.* Normally due to dissolved solids or gases

which may be natural in origin, e.g., from algae or other micro-organisms, or man-made, e.g., phenol from gas liquors and tar spraying.

(d) *Temperature*. This has an influence on taste since a warm water tastes flat and insipid, partly as a result of the decreased solubility of oxygen and carbon dioxide at elevated temperatures.

CHEMICAL CHARACTERISTICS

The chemical properties of a water are usually only apparent in the results of analysis but they have a considerable influence on the suitability of the water for consumption, to support fish or for industrial purposes. In many ways, chemical properties can be more readily measured than some of the physical characteristics, like colour, which are rather subjective in nature.

Important chemical characteristics are as follows.

(a) *pH*. A measure of the intensity of acidity or alkalinity. Water is a weakly ionized substance: $H_2O \rightleftharpoons H^+ + OH^-$ and the concentrations of the ions denoted by $[H^+]$ and $[OH^-]$ have a constant product, $[H^+][OH^-] = 1 \cdot 01 \times 10^{-14}$ mole/l at 25°C (a mole is the molecular weight in grams, 17 for water). If acid is added to a water $[H^+]$ increases and $[OH^-]$ must decrease to maintain the constant product. Conversely, if an alkali is added $[OH^-]$ increases and $[H^+]$ must decrease. The concentration of hydrogen ions is expressed by $\log_{10}(1/[H^+])$ which results in a range of values from 0 to 14. A neutral solution has a pH of 7, below 7 is acid and above 7 is alkaline. Many chemical and biological reactions are controlled by pH and a fairly narrow range of pH 6–8 is usually found in water samples although upland waters are sometimes acidic with pH down to 4. Industrial wastes may have pH values as low as 1 or as high as 13 and often require neutralization before discharge.

(b) *Alkalinity and acidity*. Quantitative parameters necessary to enable calculation of the amounts of reagents required to change the pH of a sample to some new value. Alkalinity is due to the presence of carbonates, bicarbonates and hydroxides in various combinations. Most of the natural alkalinity in water is due to calcium bicarbonate which is produced by the action of groundwater and carbon dioxide, from soil micro-organisms, on limestone and chalk.

$$\underset{\text{insoluble rock}}{CaCO_3 + H_2O + CO_2} = \underset{\text{soluble}}{Ca(HCO_3)_2}$$

The alkalinity in water is useful in that it acts as a buffer against changes in pH.

Acidity in natural waters is largely due to carbon dioxide which will not produce a pH lower than 4·5. Mineral acids from industrial wastes can give much lower pH values. Because of the complex nature of the carbon dioxide/bicarbonate system, alkalinity can exist down to pH 4·5 and acidity can exist up to pH 8·2.

(c) *Hardness.* Due to the presence of metallic ions, basically calcium (Ca^{++}), magnesium (Mg^{++}) and iron (Fe^{++}) which possess the properties of preventing lather formation with soap and forming scale in hot water systems. These metals are usually associated with bicarbonates, sulphates, chlorides and nitrates. Hard waters sometimes have a pleasant taste, e.g., from chalk strata, but have the economic disadvantages of increased soap consumption (although they do not affect the use of synthetic detergents) less effective laundering, and higher fuel and maintenance costs in hot water systems because of the scaling they produce.

(d) *Dissolved oxygen (DO).* Oxygen is a vital element in water since without it only the lowest forms of life can survive. Unfortunately, oxygen is only slightly soluble in water (9·1 mg/l at 20°C) and fish, such as trout and salmon, require at least 5 mg/l DO to live satisfactorily although some coarse fish can exist at DO levels of 2 mg/l. Pollution of water by organic matter rapidly utilizes the DO by biological oxidation and thus the receiving water may become depleted of oxygen with consequent death of fish. Waters saturated with DO have a pleasant taste whereas a water with no DO, e.g., boiled water, tastes flat and insipid. Oxygen is removed from boiler feed waters because its presence encourages corrosion.

(e) *Oxygen demand.* The importance of oxygen in water makes it essential to be able to measure the amount of oxygen required to oxidize impurities in the water. Most of the oxygen demand is due to the oxidation of organic matter by micro-organisms but inorganic reducing agents, e.g., hydrogen sulphide, also consume oxygen. Various parameters are used to assess the oxygen demand of a sample;

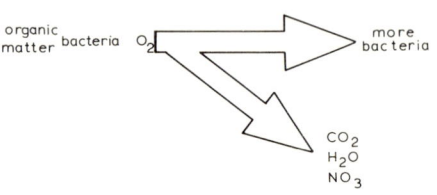

FIGURE 1. Aerobic oxidation

(i) *Biochemical Oxygen Demand (BOD)* which is a measure of the oxygen required by micro-organisms to oxidize organic matter in the presence of free oxygen, i.e., under aerobic conditions (Fig. 1), when the end products are carbon dioxide and water, sulphate, nitrate and phosphate; the standard period of test is five days at 20°C. In the absence of free oxygen, biological oxidation takes place much more slowly anaerobically (Fig. 2) with unstable and unpleasant end products like hydrogen sulphide and methane.

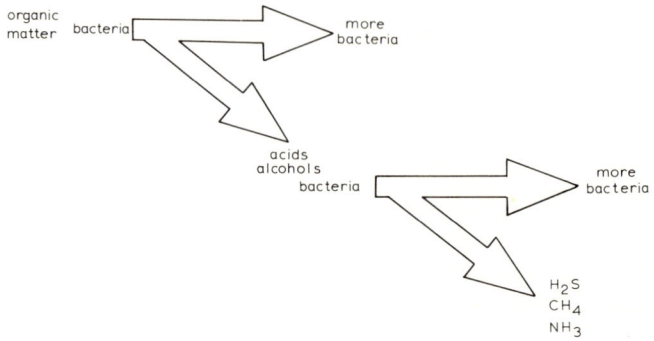

FIGURE 2. Anaerobic oxidation

(ii) *Permanganate Value (PV)*, a chemical oxidation using acid potassium permanganate solution for four hours at 27°C.

(iii) *Chemical Oxygen Demand (COD)*, which is a more vigorous chemical oxidation using boiling potassium dichromate and concentrated sulphuric acid for two hours. COD values are greater than BOD values which are normally greater than the PV figure. The COD and PV determinations give results after a few hours but because of the slow rate of biological oxidation the BOD test requires a period of five days and, even then, by no means all of the organic matter will have been stabilized in most samples.

(*f*) *Nutrients*. Materials such as nitrogen and phosphorus compounds which play an important part in biological growth and are thus necessary for efficient biological treatment of wastes. About one part of nitrogen is necessary for every 20 parts of BOD and one part of phosphorus is required for every 120 parts of BOD. The presence of nutrients in their oxidized states, as nitrates and phosphates, is responsible for accelerating alterations in the quality of lake waters, the process of eutrophication, so that the existence of nutrients in effluents discharged to such waters sometimes gives cause for concern.

(*g*) *Chloride*. Responsible for salty tastes in water which, at concentrations of chloride much in excess of 1000 mg/l, make the water undrinkable. Chlorides are also good indicators of sewage pollution since there is a high chloride content in urine (an adult discharges about 6 g of chloride each day).

When industrial wastes are present many other chemical characteristics may be of interest, e.g., grease, heavy metals, cyanides, etc.

BIOLOGICAL CHARACTERISTICS

All natural waters and most wastewaters contain a variety of living micro-organisms and a surfacewater not subject to heavy pollution supports a complex ecological system (Fig. 3). Micro-organisms are responsible for several serious waterborne diseases and, thus, the presence of such pathogenic micro-organisms in water is something to be avoided. On the other hand, most micro-organisms are harmless to man and, indeed, drinking water is not normally sterile although it has been disinfected to kill pathogenic species. Micro-organisms are capable of stabilizing organic matter and this property is utilized by nature in the self-purification of polluted waters and by man in the more intensive activity of a wastewater treatment plant.

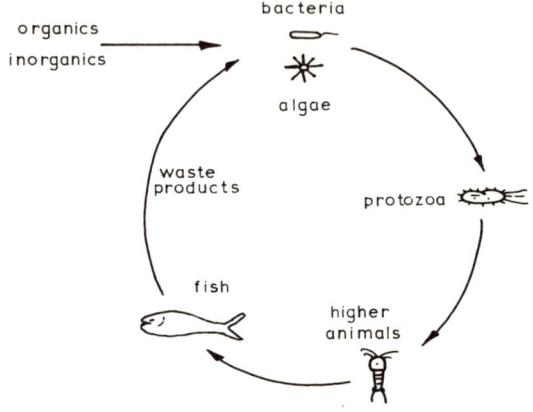

FIGURE 3. The biological cycle in surfacewater

In order to appreciate the part played by micro-organisms in water quality it is necessary to understand a little of how they live. Two basic forms of metabolism are found depending on whether or not the organisms require an external source of organic matter. Autotrophic organisms are able to synthesize their organic matter from inorganic compounds, e.g., most plants use light energy to produce organic matter, i.e., plant growth, and oxygen in a photosynthetic reaction. Heterotrophic organisms must have ready-made organic matter in their food. Some micro-organisms (known as aerobes) require free oxygen, others (anaerobes) can exist in the absence of oxygen whilst some, known as facultative forms, can exist aerobically or anaerobically depending upon the availability of oxygen in their environment. Micro-organisms, which range in size from 0·1 μm to 1 mm, are plants or animals and Table 1 shows the main features of the types important in water. Plants require soluble

Nature and Occurrence of Water

food which may be either organic or inorganic depending on the organism but animals must have organic food. Although the smallest animals can exist on soluble food alone most of the microscopic animals eat solid food, e.g., protozoa eat bacteria. Such food chains are nature's way of controlling the growth of biological populations.

Table 1. *Types of micro-organisms*

Characteristic	Plants				Animals		
	Viruses	Bacteria	Fungi	Algae	Protozoa	Rotifers	Crustaceans
Microscopic	×	×	×	×	×	×	
Macroscopic				×			×
Inorganic food only		× (some)		×			
Organic food only	×	× (some)			×	×	×
Require oxygen		× (some)	×		×	×	×
Photosynthetic		× (some)		×			
Pathogenic	×	× (some)	× (some)		× (some)		

The naming of biological organisms is complicated but this is inevitable because of the vast number of different organisms. A specific type of organism is denoted by its species name and a collection of species with similar characteristics is given a generic name, e.g., *Vibrio cholerae*, a member of the *Vibrio* genus, in fact the particular member responsible for cholera.

Owing to their small size, identification and counting of microorganisms is not possible with the naked eye. An optical microscope which has a maximum magnification of up to about 1000 times must be used and, to aid viewing, the micro-organisms are usually stained with dyes to make them contrast with the background (Plate 1). The smallest organisms cannot be identified by their physical features alone so that knowledge of their biochemical properties must also be used to make positive identifications possible. The larger organisms, like algae, can be identified visually and numbers can be assessed in special counting cells which hold a known volume of sample under the microscope. Such a method is not feasible for bacteria because of their small size and estimates of the number of living bacteria in a sample are obtained from plate counts on nutrient agar medium containing the necessary elements for growth. A portion of the sample is mixed with molten agar at a temperature of about 45°C and poured onto a dish where the mixture solidifies. The bacteria can obtain food from the surrounding jelly but are unable to move about so that after an appropriate incubation period (72 h at 22°C for natural water bacteria, 24 h at 37°C for bacteria from man and other animals) colonies of bacteria appear where a single organism had been engulfed in the medium as it set (Plate 2). A count of these colonies, which are visible to the naked eye, gives an estimate of the number of bacteria originally present in the sample although no single medium and temperature will permit the growth of all the micro-organisms present in the sample.

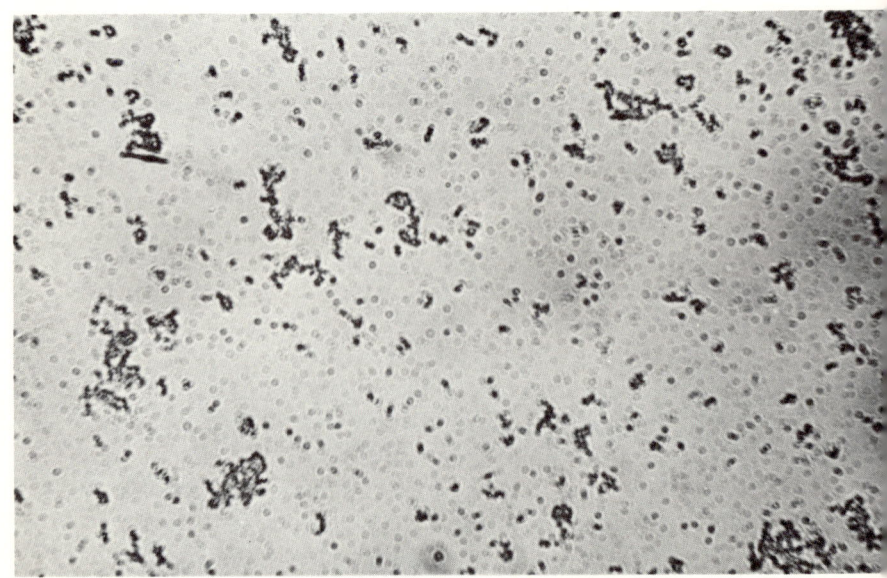

PLATE 1. Photograph of bacteria which have been treated with a dye to make them more readily visible. Magnification approximately 800 times

Nature and Occurrence of Water 13

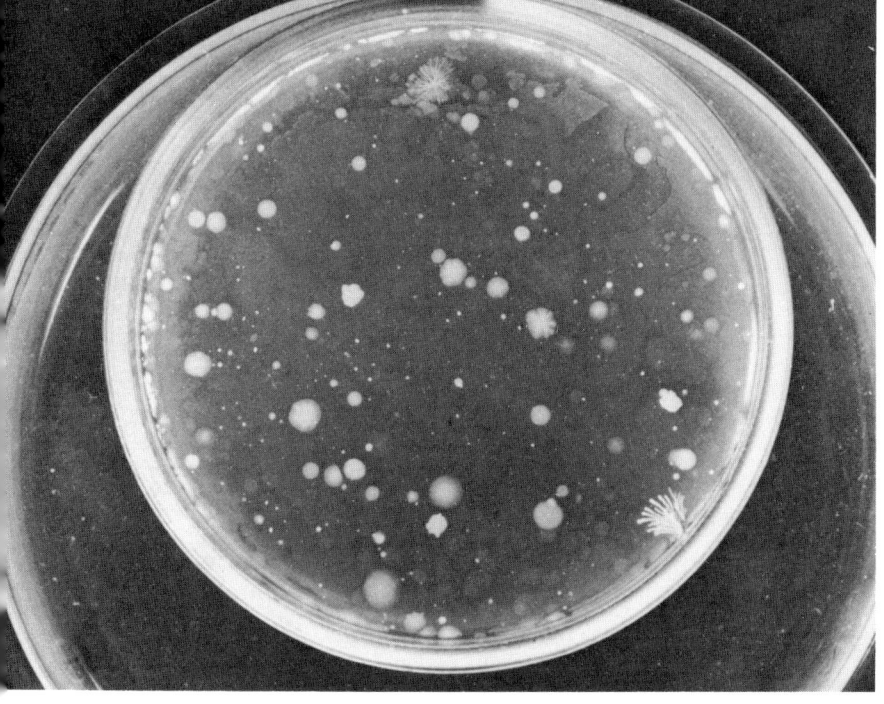

PLATE 2. Colonies of bacteria grown on nutrient agar medium. Reproduced about half full size

In order to determine the presence of a particular species of bacteria use is often made of its characteristic biochemical properties as in the multiple tube method for the estimation of *Escherichia coli*, the organism which is used as an indicator of human pollution. Coliform bacteria alone, will ferment lactose (a sugar) producing an acid and a gas. Thus, serial dilutions of the sample, e.g., 1 in 10, 1 in 100 and 1 in 1000, are inoculated into tubes of MacConkey Broth (a lactose medium) containing small inverted glass tubes. The appearance of acid, shown by an indicator which changes colour, and gas which collects in the inverted tubes, after 24 h incubation at 37°C is positive for the presence of coliform organisms (Plate 3). To confirm the presence of *E. coli* positive tubes are subcultured into fresh MacConkey tubes and incubated for 24 h at 44°C, any positive reactions being due to *E. coli*. The application of statistical analysis to the results enables an estimate to be made of the number of *E. coli* present in the sample.

Table 2 shows typical characteristics of water from various sources.

Table 2. Typical characteristics of water from various sources

Characteristic	Upland stream	Low-land river	Industrialized river	Tap water	Raw sewage	30:20 standard sewage effluent
pH units	6	7·5	8	7	6·5	7
Colour °H	70	30	40	5		60
Turbidity	10	30	50	1		30
BOD	1	3	15	1	350	20
PV	1	2	10	1	100	15
Chloride	5	50	250	50	200	200
Total solids	30	100	750	300	700	600
Coliform MPN/100 ml	500	10^4	5×10^5	0	10^7	10^5

(*analyses in mg/l except where noted*)

THE HYDROLOGICAL CYCLE

The total volume of water present in the earth's hydrosphere, i.e., in the atmosphere, on the surface and in the crust is thought to be about 10^{18} m^3 which is equivalent to about seven per cent of the mass of the earth. The salt-water in oceans and seas accounts for about 97 per cent of the total but, even allowing for this, there is about $2 \cdot 5 \times 10^6$ m^3 of fresh water for each person on the earth. This figure does not, however, represent the water which is readily available and which is probably only about one per cent of the total fresh water in the hydrosphere. The available water is largely derived from precipitation on the land which occurs as part of the hydrological cycle (Fig. 4). When rain or snow falls on the land some appears as surface run-off in streams and rivers, some is lost by evaporation from the ground and

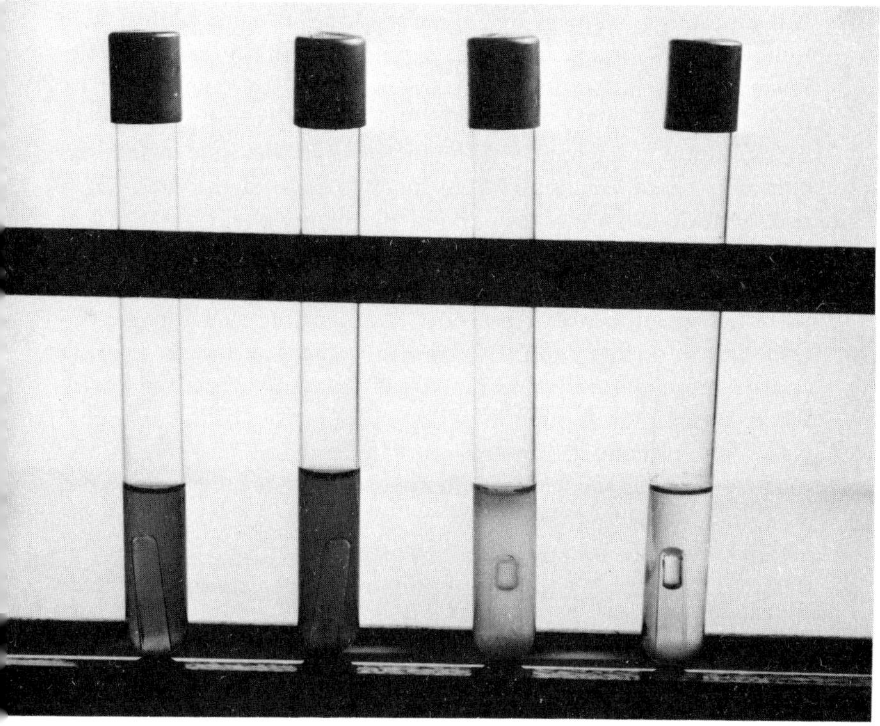

Nature and Occurrence of Water

PLATE 3. MacConkey Broth tubes for coliform counting. The two tubes on the left show no reaction but the tubes on the right have a gas bubble visible in the small tubes and a cloudy appearance (the colour of the broth has changed from purple to yellow) indicating the presence of coliform bacteria in the right-hand tubes

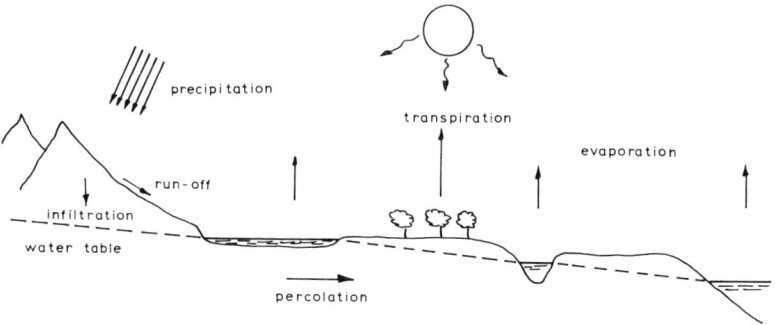

FIGURE 4. The hydrological cycle

water surfaces, some is lost by transpiration from plants. The remainder infiltrates into the ground eventually reaching the water-table where movement towards the sea takes place by percolation through the strata. The relative proportions of precipitation following the different paths in the cycle depend on climate, vegetation, topography and geology of the area. As a rough guide in temperate parts of the world about one third of the precipitation appears as surface run-off, a further third is lost by evaporation and transpiration and the remaining third replenishes groundwater reservoirs. Thus, water can only be obtained from surface run-off or by abstraction from water-bearing strata known as aquifers. In densely populated parts of the world, where the rainfall is often low, large proportions of the available water are already being utilized. In England and Wales the present demand for water is about 20 per cent of the estimated mean available resources and in areas like S.E. England the locally available resources are almost fully utilized. Since water demand is increasing at about four per cent per annum it is clear that water conservation is essential.

3 Water Use

The main uses of water are, in alphabetical order:
- (*a*) Amenity.
- (*b*) Domestic water supply*.
- (*c*) Effluent disposal*.
- (*d*) Fishing.
- (*e*) Industrial water supply*.
- (*f*) Irrigation†.
- (*g*) Navigation.
- (*h*) Power production.
- (*i*) Recreation*.

The aim of modern water resources development is to achieve multi-purpose use of water so that as many as possible of the above uses are satisfied by each portion of the water resources of an area. It will be appreciated, however, that not all the uses are compatible as far as quality and quantity are concerned. The commonest water uses are for domestic and industrial water supply and waste disposal and it is these uses which will be considered in detail.

DOMESTIC WATER SUPPLY

Man can satisfy his requirements for water intake with a daily consumption of about two litres but as higher standards of living are adopted so the demand for water increases. Piped water schemes inevitably result in the installation of water-borne waste disposal systems. Washing machines, air conditioners, garbage grinders, car washing and garden watering all add to the demand

---

* Use lowers quality of water in system or returned to the system.
† Consumptive use, water is lost from system and does not become available for re-use.

for water. Furthermore, in the complex water distribution system necessary to serve the needs of a city a degree of leakage and waste must be expected. Thus, in the UK an average daily domestic water demand is likely to be about 150 to 180 l/person day made up as shown in Table 3. In warmer countries the domestic demand may rise to as much as 500 l/person day due largely to increased demands for garden irrigation, air conditioning, baths and showers.

Table 3. Domestic water consumption in the UK (Modern Housing 1970)

Use	Litres/person day
Drinking & cooking	5
Dishwashing, etc.	15
Laundry	15
Personal hygiene	50
Closet flushing	50
Garden use, etc.	5
Waste in distribution	20
Total	160

In the UK, the domestic demand for water appears to be increasing at a rate of between two and four per cent annually which, if continued, will mean that the demand for water will have doubled by the end of the century. This growth in demand is due to two factors: the increase in population, and the increase in *per capita* consumption of water associated with a higher standard of living. There seems little likelihood of an overall reduction in population in the future although the rate of growth may, perhaps, have to be controlled. It does, however, seem likely that the *per capita* demand for water must reach a ceiling when all possible household requirements for water have been satisfied. Indeed, it is difficult to foresee how further domestic demands can arise when every household has its full complement of washing machines, dishwashers, garbage grinders and the other appliances provided by modern technology. Thus, a maximum domestic demand of about 300 l/person day seems probable in this country.

Considerable amounts of water are used thoughtlessly in the home because of its low cost and the method of charging for its consumption. Almost all water undertakings in the UK supply domestic water at a flat rate charge based on the size of the house and no account is taken of the amount of water actually consumed. There is thus no incentive to prevent waste, although legislation is in force to prevent waste and many water undertakings offer tap washers and fit them free of charge in attempts to reduce waste. The adoption of metering for domestic water supplies is often suggested as a method of reducing water consumption, particularly if a sliding scale tariff were used to discourage excessive water use. Evidence from countries which practise metering is, however, conflicting and it would clearly be undesirable to restrict the consumption of water by economic measures to the extent that hygiene suffered. Attempts to reduce domestic consumption have so far been largely based on the installation of dual-flush wc cisterns giving a small flush for use after urination with the conventional flush reserved for use after defecation. Such cisterns are in use in various parts of the country and have been found to reduce domestic water consumption by about 10 per cent. The more general adoption of showers in place of baths would also make a significant reduction in the amount of water used for domestic purposes. Careful design of water plumbing systems to avoid long pipe runs would reduce the amount of hot water wastage caused by allowing taps to run until hot water reaches them.

Considering a single dwelling, the demand for water varies over a wide range throughout the day and in small communities this effect is very marked. The fluctuation in demand decreases with increasing size of the supply population because of the varying patterns of the inhabitant's lives and the balancing effect of a large distribution system.

Water undertakings must provide a wholesome supply and, whilst such a supply has not been legally defined, a water would be likely to satisfy statutory requirements if it were free from the following impurities.

(a) Visible suspended matter.
(b) Colour.
(c) Taste and odour.
(d) Objectionable dissolved matter.
(e) Bacteria indicative of pollution.
(f) Aggressive constituents.

The water should be fit for human consumption, i.e., potable, and it should also be aesthetically attractive, i.e., palatable. Additionally, it should be suitable for other domestic uses, such as washing, and should not, therefore, contain excessive hardness.

In determining quality standards for drinking water it is necessary to consider many characteristics. Such properties as colour and taste may be limited purely on aesthetic grounds, e.g., the natural golden-yellow colour of upland catchment supplies is due to humic and tannic acids (akin to tea) which are not hazardous to health at the concentrations found in natural waters. Nevertheless, the colour they produce may well be objectionable to many consumers and for this reason colour is removed as far as possible for large supplies. With supplies for a small community the removal of colour from an otherwise satisfactory water would often be uneconomic. Thus, many villages in western Scotland have highly coloured supplies which are, nevertheless, quite safe to drink.

When considering poisonous materials such as arsenic, cyanide and heavy metals it is possible to use medical evidence of toxicity to help in determining maximum allowable concentrations. This is not easy, however, because the level adopted must allow for the fact that a person may consume water containing these impurities every day of his life and, thus, the effects of long term exposure to sub-toxic concentrations must be considered. With many of the newer chemicals found in industrial wastes and used in agriculture, information on possible long term hazards is not yet available so that 'safe' concentrations may have to be formulated on the basis of incomplete information.

Materials such as chlorides and nitrogen compounds are not necessarily harmful in themselves but indicate the presence of contamination which is potentially harmful. The most sensitive indication of water quality is given by bacteriological analysis.

Water Use

Polluted drinking water carries a considerable danger of spreading waterborne disease by transporting the pathogenic bacteria which are the causative agents of the diseases. The presence of such micro-organisms can be demonstrated by specialized techniques but a much more rapid and sensitive measure of water quality is obtained by using the indicator organism *Escherichia coli*. This bacteria is normally present in the human intestine and is excreted in large numbers and would therefore be present in sewage or polluted water in much higher numbers than the pathogenic bacteria. The pathogenic bacteria are also likely to die more rapidly than *E. coli* in the unfavourable environment outside the body. The recommended standard of bacteriological quality for drinking waters in the UK is that *E. coli* should not be present in 100 ml of water. At this level it is statistically highly unlikely that any pathogenic bacteria would be present and in any case it is known that more than one organism must be ingested before infection occurs. Some recent work in the USA has shown that when 20 volunteers ingested 1000 viable typhoid bacteria no cases of typhoid resulted and even the ingestion of 10 million viable bacteria produced typhoid in only half of those exposed. The presence of viruses in polluted waters is a cause of some concern because they are not so readily destroyed by conventional methods of water treatment as are the bacteria and because the number which must be ingested before infection occurs is relatively low.

In this country there are no mandatory water quality standards for potable supplies, each case being considered on its merits within the general guide lines of the standards formulated by the World Health Organization which are shown in Table 4.

INDUSTRIAL WATER SUPPLY

The industrial demand for water can be a significant proportion of the total demand for water in large urban areas, e.g., in Birmingham the industrial demand is somewhat larger than the domestic requirement. Thus, industrial demand for water is likely to play an important part in future water requirements.

Tabel 4. World Health Organization drinking water standards
(Some Examples of Recommended Concentrations)
International Standards 1963

Characteristic	Limit of general acceptability	Allowable limit
Total solids	500	1500
Colour (°H)	5	50
Turbidity	5	25
Chloride	200	600
Iron	0·3	1
Manganese	0·1	0·5
Copper	1·0	1·5
Zinc	5	15
Calcium	75	200
Magnesium	50	150
Magnesium and sodium sulphate	500	1000
Nitrate (as NO_3)	45	—
Phenols	0·001	0·002
Synthetic detergents (ABS)	0·5	1·0
Carbon-chloroform extract	0·2	0·5
pH	7–8	min 6·5 max 9·2

European Standards 1970

Toxic substances maximum concentration		Concentration above which trouble may arise
Lead	0·1	Fluoride 1·0–1·7 (Fluorosis) (depends on climate)
Arsenic	0·05	Nitrate 100 (Methaemoglobinaemia)
Selenium	0·01	Iron 0·1 (Taste and staining)
Chromium (6+)	0·05	Chloride 200 (Taste)
Cadmium	0·01	Sulphate 250 (Gastrointestinal irritation)
Cyanide	0·05	Zinc 5 (Taste)

(Concentrations in mg/l except where noted)

In 95 per cent of samples examined throughout a year coliform bacteria should be absent.

Water Use

Most existing industrial supplies are already metered so that more precise demand data are available than for domestic supplies but many small industries have their own independent sources, often boreholes or wells. Future demands from such industries, however, may well have to be met from public supplies. Although it is known that metered industrial demand has been growing at about 2·5 per cent annually over the past few years it is not possible to predict future demands with great confidence. Another factor which makes predictions difficult is the rapidity with which industrial demand can change. A new factory in a community may require large supplies of water to be made available within a short period of time. On the other hand, changes in processes or closure of factories due to rationalization of industry may suddenly result in a large fall in water demand leaving the water undertaking with expensive plant and facilities for which it has no immediate use.

If water is cheap its industrial use is often wasteful, e.g., taps and hoses are left running and processes may include inefficient washing procedures. As water becomes more expensive, economies in its use become more attractive. In addition, much of the water consumed industrially is finally discharged as industrial effluent. Charges for the treatment of such effluent usually contain a charge for volume, so this, too, may influence the industrialist to introduce economies in water use. Because of the many factors which affect industrial water demands, it is difficult to give specific figures for the demands in a particular industry but Table 5 shows typical water consumptions for a number of processes.

Probably the largest single use of water in industry is that of cooling, since many processes require the removal of heat from a product or during a reaction. Condenser cooling-water amounts to almost the whole water demand in power stations and much of the water required in the chemical industry is used in condensing and cooling operations. There are three types of cooling system as shown in Fig. 5. In the 'once-through' system, the water is returned to the original source after one pass through the condensers. Very large quantities are required in this method and a considerable amount of heat is transferred to the source. Such thermal pollution can have serious effects on the environment since most biochemical and chemical reactions are temperature dependent. The 'open recirculation' system is much more economical in water usage since the water is recycled in the plant and is itself cooled by evaporation in spray ponds or cooling towers. Some water is lost by evaporation (usually about two per cent) so that make-up water is required but this is small compared with the volume of water in circuit. For a 2000 MW power station a 'once-through' cooling system would require a flow of about 50 m^3/s, whereas on an 'open recirculation' basis with 20 000 m^3 in circuit the make-up water would only be about 0·5 m^3/s. A 'closed recirculation' system, of which the motor vehicle cooling system is the commonest example, is sometimes used for industrial purposes and, in the absence of leakage, there is no further demand for water once the system has been initially filled. Certain industries use water for direct cooling, e.g., quenching of coke. This method evaporates large amounts of the water and leaves the remainder in a contaminated state.

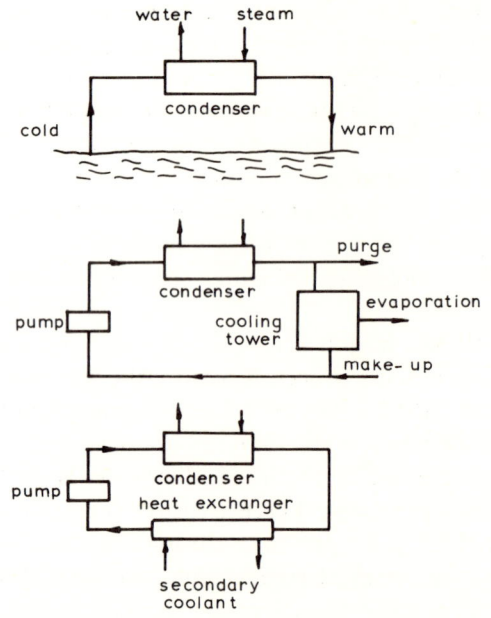

FIGURE 5. Types of cooling systems: once-through, open recirculation and closed recirculation

Table 5. Typical industrial water demands

Industry	Water demand m³/tonne product
Baking	4
Chemicals	up to 1100
Coal-mining	5
Laundering	45
Milk-processing	4
Paper-making	90
Steel production	45
Sugar-refining	8
Synthetic fibres	140
Vegetable-canning	10

Steam-raising plant used for power generation and heating purposes uses relatively little water as long as most of the steam is returned as condensate. Some processes may, however, use steam in these operations and in such cases water is lost from the system. Water may also be incorporated in the product or used in the actual manufacturing process. In the food and beverage industries much of the product may be water and large amounts of water are needed for cleansing utensils and plant. In the chemical industry, water is a major constituent of many products and, here again, rinsing operations consume large volumes of water. Certain industries use water to transport materials between processing operations, e.g., sugar beet refining where the beets are suspended in water conveyed around the factory in flumes. The paper-making industry also uses considerable amounts of transport water. Materials such as china clay or boiler ash are often slurried with water to produce a suspension which can be readily pumped.

The quality of water required for industrial purposes is as variable as the quantity. Although many industries use potable supplies obtained from water undertakings, water of such quality is by no means always necessary. While potable water is cheap and plentiful, however, and must be supplied to the premises for human consumption, the provision of an alternative supply of lower-grade water for industrial process use may be difficult to justify. There are many industrial uses of water, e.g., cooling purposes, which could be, and indeed are already, satisfied by non-potable supplies from polluted rivers or sewage effluent from municipal treatment plants. The industrial estate at Warrington, Lancashire is provided with a non-potable supply of partially-treated river water for process use as well as the normal potable supply. Provided care is taken to prevent cross-connection between the two supplies, which could be a hazard to health, such a service plays a useful part in water conservation. It is an axiom of water conservation that water of a quality higher than necessary for a particular use should not be consumed by that use.

Certain industrial uses of water may require a supply of very high quality, e.g., feed water for high pressure boilers must contain less than 1 mg/l of impurities. Demineralized water is used in such circumstances. Although very pure, such water would be quite unacceptable for domestic purposes because of the absence of dissolved salts and gases which give water a pleasant taste. Other industries requiring high-purity waters include the manufacture of pharmaceuticals and semi-conductors. The manufacture of high-grade art papers must have a water of lower colour than would be quite acceptable for potable supply. Table 6 gives some examples of water quality requirements for various industrial purposes.

Table 6. Typical quality requirements for industrial process water

Process	Maximum desirable concentration mg/l				
	Total solids	Alkalinity	Turbidity	Colour	Iron & manganese
Boiler feed:					
low pressure	250				
high pressure	1	0	0	0	0
Dyeing	50	20	5	5	0·2
Paper (art)	250	50	5	5	0·05
Semi-conductors	1	0	0	0	0
Soft drinks	250	50	2	2	0·2
Synthetic fibres	700	500	5	5	0

Water Use

WASTEWATER

Once water has been provided for domestic or industrial use the inevitable consequence is the production of wastewater which normally requires treatment before it can be safely released to the environment. Almost all of the water supplied to domestic consumers in temperate climates is returned as sewage, losses of water due to evaporation and garden watering being more or less counterbalanced by infiltration of groundwater into sewers and drains. Even in hot countries, where much of the domestic water is used for irrigation, about two thirds of the water appears as sewage. In a similar manner much of the water supplied to industrial consumers reappears as effluents although evaporation and incorporation in the product can reduce the proportion returned. Any factory which consumes significant volumes of water and claims to produce no effluent should be viewed with suspicion.

Domestic sewage contains about 1000 mg/l of impurities of which about two thirds are organic in origin. About half of the impurities are present as suspended solids, the remainder being in solution. The main organic compounds present in domestic sewage are: nitrogenous compounds—proteins and urea; carbohydrates—sugars, starches and cellulose; fats—soap, cooking oils and greases. Inorganic constituents include chloride, metallic salts and road grit if a combined sewerage system is used. Dissolved gases may be present, fresh sewage contains dissolved oxygen but this is rapidly lost due to oxidation of the organic matter and stored sewage soon becomes septic evolving hydrogen sulphide. In hot climates the sewage may turn septic in the sewers and the hydrogen sulphide produced can result in the formation of sulphuric acid which attacks the sewer structure. Hydrogen sulphide is a highly toxic gas and its presence in sewage is a hazard to workers in sewers. Sewage contains large numbers of bacteria, about 1 million/ml and other microorganisms are also found in smaller numbers.

The nature of industrial wastewaters is as variable as the number of process from which they arise (Table 7). Cooling water discharges are basically unchanged in composition except for the rise in temperature and the consequent reduction in dissolved oxygen, which can be a serious problem on its own. On the other hand, wastewaters from dairies and other food processing establishments are many times worse than domestic sewage. Farm effluents are often highly polluting and the connection of a farm to a small rural sewage works may have disastrous consequences due to overloading. The metallic ions such as chromium, nickel and zinc found in metal finishing wastes are toxic to biological systems and their discharge untreated to surface waters can result in the wholesale killing of fish and eventual fishless rivers of which there are several examples in industrial areas.

Table 7. Typical wastewater characteristics
(results in mg/l except for pH)

Waste	BOD	SS	Total solids	Alkalinity	pH
Domestic sewage	300	400	700	250	6·5
Cotton kiering	1500				11
Dairy	1000		2000		
Distillery	20 000		50 000		
Laundry	2000		4000	500	10
Silage liquor	60 000				
Sugar beet	500	5000	6000		

4 Collection of Water

In simple civilizations, communities grow up close to water sources such as a lake, river or spring. As development occurs the demand for water outstrips the locally available supply so that other sources must be brought into use. In addition, the waste-products of the community must be treated and discharged in such a manner that the water sources do not become contaminated.

The conventional sources of water are:

(*a*) Upland catchments.
(*b*) Rivers.
(*c*) Groundwater reservoirs.

In areas of water shortage, treated wastewater and seawater may be used to supplement more conventional sources. Knowledge of the reliable yield of any source is essential for design purposes and for the efficient operation of water supply schemes. Conventional sources of water depend on replenishment by rainfall and, thus, their yield is influenced by the random nature of precipitation. The yield is further influenced by factors such as temperature, vegetation, topography, etc. By statistical analysis of past records of rainfall and run-off it is possible to determine the probability of obtaining a given yield from a particular source or, more commonly, to determine the yield which could be obtained with a given probability of failure. Groundwater sources are more difficult to analyse because of the problems in obtaining sufficient information about the properties of the aquifer. To enable yield predictions to be made it is essential to have available data from a hydrometric scheme set up to measure the important parameters in the hydrological cycle at a number of key points in the catchment or aquifer. The longer the period for which such data is available the more reliable the predictions are likely to be.

HYDROLOGICAL MEASUREMENTS

RAINFALL MEASUREMENT

Rainfall varies widely in different parts of the world, from deserts in Africa where it is almost unknown, to mountainous areas like the Himalayas where 15 m a year have been recorded (rainfall is measured as the depth of water which would collect over the area if there were no loss by run-off, infiltration or evaporation).

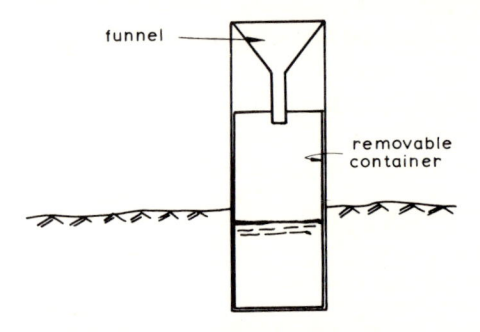

FIGURE 6. The standard rain gauge

In the UK, the long-term average annual rainfall is 1050 mm, ranging from 500 mm/year in parts of East Anglia to 5000 mm/year in parts of western Scotland. About half of England receives less than 750 mm/year and about one sixth of the British Isles receives more than 1250 mm/year. Rainfall has long been a subject of interest to scientific observers and, since 1860, daily rainfall records from several thousand gauges all over the country have been published annually in *British Rainfall*. The standard copper rain gauge (Fig. 6) is emptied at 9 am each day and the volume of rainfall collected in the previous 24 hours is measured in a special cylinder graduated in millimeters of rainfall. Daily gauges tend, of course, to be situated in the more populous parts of the country and their use in isolated areas is limited by the availability of observers such as shepherds who live close to the gauge site. Daily gauges give no indication of the intensity of rainfall over short periods which is of particular concern when the possibility of flooding is under investigation. Much

Collection of Water

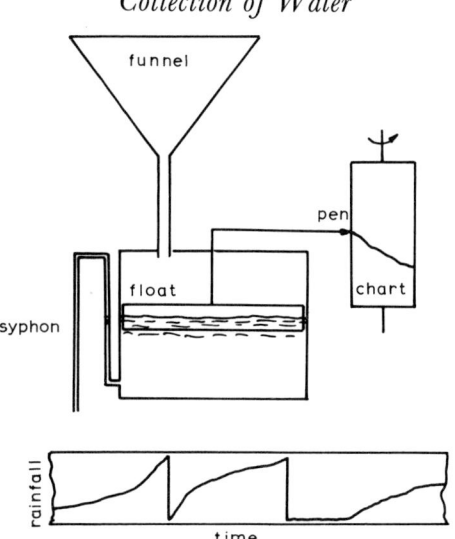

FIGURE 7. One type of recording rain gauge. This produces a 30-day record on a strip chart, giving both intensity of rainfall and total precipitation. When full the float chamber empties via the syphon tube thus giving good accuracy for small amounts of rain while still being capable of recording high rainfalls

use is now made of recording gauges (Fig. 7) which require attention at monthly intervals only and provide records of intensity and total rainfall. Remote gauges of this type may be linked to a central station by some form of telemetry to provide information for flood warning schemes and for the operation of control works.

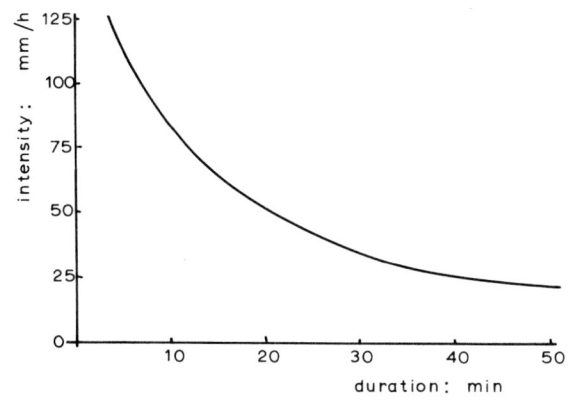

FIGURE 8. A typical rainfall intensity–duration curve for a return period of 1 year

A characteristic property of rainfall is that the intensity for a given probability of occurrence or return period, decreases as the duration of the storm increases (Fig. 8). For a given duration of storm in a particular location, the intensity of rainfall depends on the return period, i.e., the likely interval between similar events, as shown in Table 8. The Bilham formula is often used for the calculation of rainfall intensities in the British Isles:

$$\text{Rainfall intensity mm/h} = \left[\frac{14 \cdot 2 F^{0 \cdot 28}}{t^{0 \cdot 72}} - \frac{2 \cdot 54}{t}\right]$$

where F = return period in years.
 t = duration of storm in hours.

Table 8. Bilham rainfall intensities (10 minute storm)

Return period	Intensity
1 year	36 mm/h
2	47
5	65
10	83
20	104
50	139
100	172

RUN-OFF MEASUREMENT

Surface run-off in streams and rivers can be measured by a number of gauging techniques all of which determine the flow indirectly.

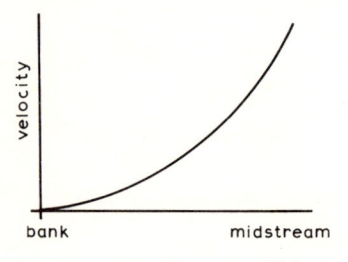

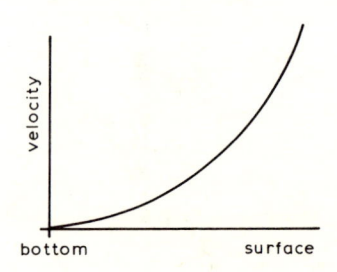

FIGURE 9. Velocity distributions in a stream channel

Collection of Water

PLATE 4. A propeller-type current meter. The meter can be fixed at various depths on the bar and the vane at the rear helps to keep the meter directed into the flow. Rotation of the propeller makes an electrical contact the impulse from which passes through the cable to a timing device

If the cross-section of the channel is accurately known the quantity flowing can be determined by measuring the mean velocity of flow. In a natural channel of irregular profile there is considerable variation in velocity both horizontally and vertically (Fig. 9), so that to obtain the mean velocity it is necessary to measure the velocity at a number of points in the flow. The velocity can be obtained by using a current meter, a device with propeller blades or small cups which revolve in the flow (Plate 4). As the shaft rotates, an electrical contact is made which actuates a buzzer or digital counter so that the time for a known number of revolutions can be recorded. Current meters are calibrated by towing them in a tank at a number of known velocities. In small streams the meter may be held on a staff, in deeper flows the meter may be suspended from a bridge or a boat. For permanent gauging stations a catenary cable is usually suspended across the river. To give accurate results the current metering must be done in a stretch of the river which is straight, with a reasonably uniform bottom, free from weeds and other obstructions. The cross-section is split into a number of panels (Fig. 10) and the velocity at the edge of each panel is measured at several depths to enable the mean velocity to be calculated. For rapid work it can be as-

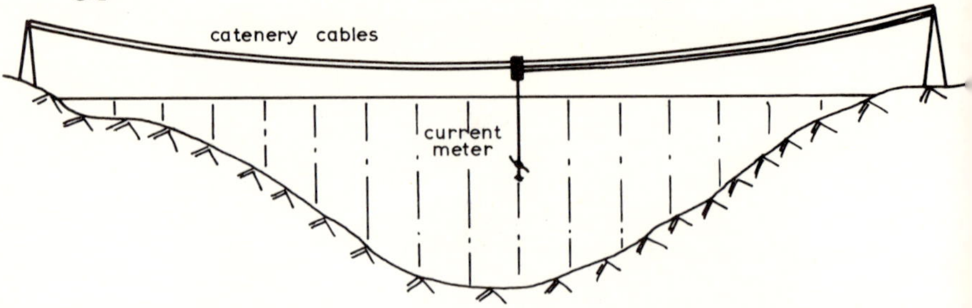

FIGURE 10. Current metering. Velocities are determined at a number of depths on each vertical to enable the flow in each panel to be calculated as accurately as possible

sumed that the velocity at 0·6 depth from the surface is equal to the mean velocity. Knowing the mean velocity and the area of each panel it is then possible to calculate the total discharge at the section. By making a number of gaugings at different river levels it is possible to produce a stage-discharge curve for that particular section so that a simple water level measurement by staff or level recorder (operated by a float) can be converted to a discharge.

Chemical dilution techniques can also be used to measure flow. This method is based on the addition to the flow to be measured of a known amount of tracer, e.g., salt solution or a dye. After mixing has taken place samples are withdrawn downstream at known time intervals and analysed for the tracer with typical results as shown in Fig. 11. By considering a quantity balance for the tracer it is possible to calculate the flow in the river, e.g.,

$$Q = \frac{q(C_0 - C_2)}{(C_2 - C_1)},$$

where Q = discharge in river;
  q = rate of tracer injection;
  C_0 = tracer concentration in injected solution;
  C_1 = background concentration of tracer;
  C_2 = tracer concentration after mixing in flow.

The method depends on the achievement of complete mixing of the tracer with the flow and this may take a considerable distance in slow moving rivers thus limiting the value of dilution gauging. In large rivers the cost of tracer needed can be considerable and

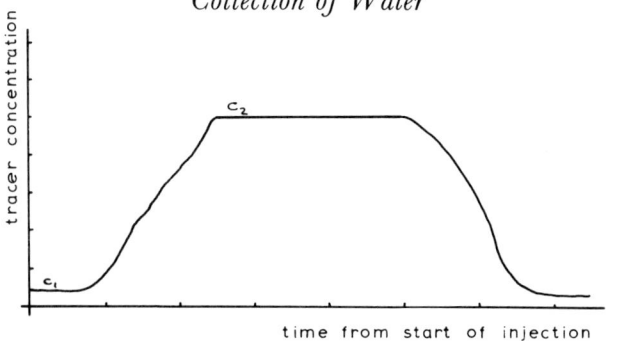

FIGURE 11. Typical analytical results from a chemical dilution gauging

care must be taken to ensure that the tracer does not significantly impair water quality.

For many purposes a continuous record of flow is desirable and, although this can be obtained from a level recorder installed at a current metering site, natural stream beds change in configuration due to scour and deposition, weed growths etc., so that the results are of limited accuracy. In situations where an obstruction to the flow can be tolerated, e.g., where the water is unnavigable, a gauging structure in the form of a notch or weir (Plate 5) will give continuous records of greater accuracy than obtainable from natural sections. Such gauging structures have constant sections and discharge relationships of the form

$$Q = Cbh^{3/2},$$

where Q = discharge;
C = a constant depending on the type of structure;
b = width of structure;
h = head on crest of structure.

Thus, a float installed in a well connected to the upstream side of the structure will measure the head over the crest and, knowing the calibration, a continuous discharge record is obtained either in chart form or as punched tape for computer input. The initial calibration of the structure is often obtained by making a scale model and observing its characteristics in the laboratory but site checks on the structure, as installed, by current metering or chemical dilution methods are advisable since shoaling upstream of the structure may alter its discharge characteristics.

PLATE 5. A Crump gauging weir with 2:1 upstream face and 5:1 downstream face; largely self-cleansing and does not greatly impede the passage of fish

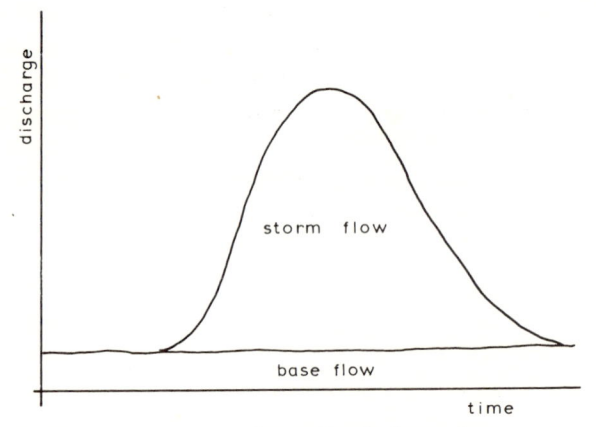

FIGURE 12. Streamflow hydrograph

A plot of flow in a river against time is called a hydrograph, a typical hydrograph (Fig. 12) indicates the dry weather, or base, flow and the effect of a storm in the catchment on the river flow. The types of hydrograph obtained from a flashy mountain stream with rapid run-off due to the steep rocky ground, and a large lowland river with an undulating rural catchment delaying run-off, are shown in Fig. 13. The area enclosed by a hydrograph represents the total volume of flow in a certain time interval. Mathematical analysis of hydrographs is of great importance in the assessment of water resources and prediction of floods and it is often useful to obtain the hydrograph in the form known as the unit hydrograph. This is based on the assumption that rainfalls of the same duration but of differing intensity will produce a hydrograph of similar form but with the flows in the ratio of the rainfalls. Thus, if a rainfall of 20 mm in 10 h produced a particular form of hydrograph a rainfall of 10 mm in 10 h would give a hydrograph the ordinates of which would be one half of those in the original hydrograph. The unit hydrograph delineates the response of the catchment for a storm of given duration. Storms of different durations have unit hydrographs of differing lengths.

When developing new water resources schemes rainfall and flow records are often available for only a short period of a few years, so interpretation is difficult. Various mathematical techniques are used to extend short term records by correlation with longer records from other catchments with similar characteristics. It is possible to generate synthetic data having the same statistical characteristics as the recorded data so that simulated flows with a particular frequency of occurrence can be generated to assess the yield of the scheme or the return period of floods of a predetermined magnitude.

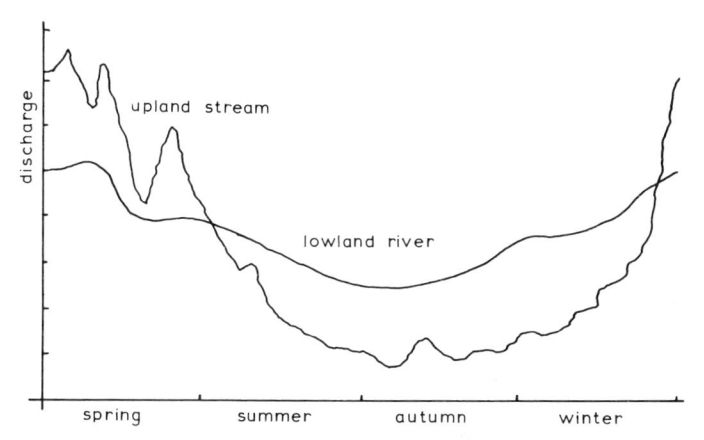

FIGURE 13. Typical hydrographs for a flashy upland stream and a large lowland river

MEASUREMENT OF EVAPORATION AND TRANSPIRATION

Evaporation from the ground and free water surfaces, and transpiration by vegetation can account for considerable amounts of water loss and, in hot climates, may consume almost all of the water available from precipitation. Serious evaporative losses take place from storage reservoirs in such areas and attempts to reduce these losses have included the spreading of thin chemical films on the water surface. Unfortunately, wind action rapidly breaks up such protective films so that their use is of limited value.

Transpiration by vegetation accounts for large volumes of water and, indeed, transpiration during the day may well remove water from the ground more rapidly than it can be replaced. Measurements of evaporation from a free water surface are made with evaporating pans which may be floating in a lake or set on the ground. Records of the water level in the pan are kept and, at the same time, measurements are made of temperature, humidity and wind speed, all of which affect the amount of evaporation. Transpiration in natural surroundings is more difficult to measure accurately. The commonest method is to use a lysimeter which is an enclosed volume of soil incorporating samples of the particular vegetation with provision for determining the precipitation on the area, the surface run-off and any groundwater flow. On a small scale a lysimeter can comprise a column of soil placed in an impermeable tube and dosed with known amounts of artificial rainfall. If the whole tube is mounted on a weighing machine the loss of weight gives an indication of evaporation and transpiration.

QUALITY MEASUREMENT

In any consideration of water resources, measurement of the quantity of water available only gives part of the picture. It is essential to have particulars of the water quality as well since these will give an indication of the suitability of the water for various uses and the degree of treatment necessary for effluent discharges or for water abstracted from the source. Information on water quality has in the past been largely based on the results of analyses of grab samples collected at intervals but such samples can be misleading and there is now increasing use of remote quality monitoring apparatus. These installations provide continuous records of certain quality parameters such as, turbidity, pH, temperature, dissolved oxygen, conductivity (as an indication of total dissolved solids) and possibly ammonia and/or nitrate. The monitors are normally sited at flow gauging stations so that correlation between quality and flow data can be obtained.

SOURCES OF WATER

As communities expand, the initial natural sources of water, rivers and lakes or springs become insufficient to satisfy the demands and some form of resource development becomes necessary. In the case of large rivers, ample water may be available for abstraction even during periods of low flow so that no control would be necessary from the point of view of quantity. By their very nature however, large rivers often tend to have fairly dense development along the banks so that considerable amounts of effluent will reach the river. Thus, in drought conditions, the proportion of effluent in the flow may be higher than desirable for treatment by conventional means and it may be necessary to construct regulation works to provide sufficient water for dilution purposes. With smaller rivers the flow is likely to be much more variable so that, as Fig. 14 shows, the reliable yield of such a source is low compared with the average flow. Except in very special circumstances it is not possible to abstract all of the water in the river for supply since flow must be maintained for the benefit of downstream users and probably to preserve fishing. Upland streams have pronounced flow variations with very low flows in dry weather and high flood flows so that, without control, most of the flow cannot be utilized for reliable water abstraction. The water from such catchments is usually of high quality requiring little treatment to produce a potable supply although colour removal can sometimes be troublesome. Most of the early development of water resources was therefore directed towards better utilization of upland catchments by impounding the storm flows which would otherwise pass straight to the sea.

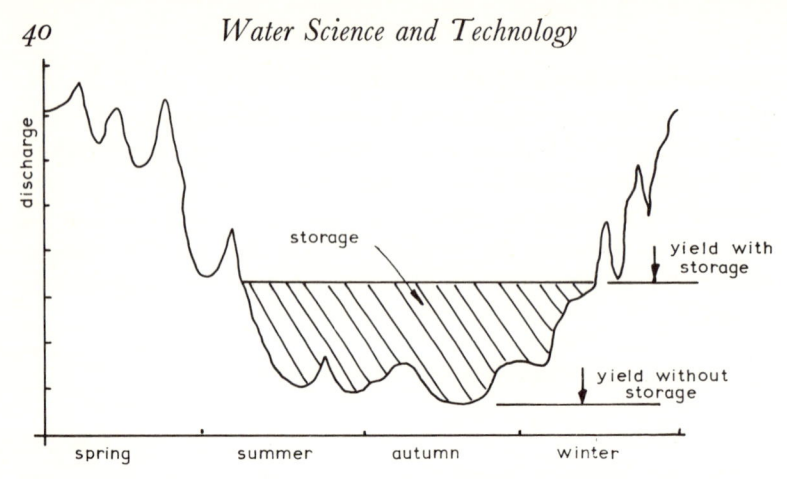

FIGURE 14. Effect of storage on yield. With a flashy stream only a low reliable yield can be provided. Storage of winter and spring flood flows can be used to produce a much larger reliable yield

IMPOUNDING RESERVOIRS

By impounding the flow of an upland stream in a suitable valley it is possible to maintain a much greater abstraction than from the same catchment without impoundment (Fig. 14). The traditional development of an upland catchment for water supply purposes involves the use of a direct supply reservoir (Fig. 15) feeding the community via an aqueduct, a suitable flow for downstream users being provided by the discharge of compensation water from the reservoir which, in the past, has often been about one third of the reservoir yield. Such schemes are typified by the reservoirs of Manchester Corporation in the Lake District and those of Liverpool Corporation and Birmingham Corporation in Wales (Plates 6 and 7). To maintain the high quality of water in such reservoirs it has, until recently, been the practice of water undertakings to restrict access to the reservoir and catchment but such measures have resulted in considerable ill feeling on the part of the public. Proposals for new impounding reservoirs almost always arouse strong objections from many groups of people who cite the severe restrictions still in force on some of the older catchments. In fact, many water authorities now allow activities like fishing and sailing on their raw water reservoirs as well as providing facilities for walking and picnicking in the

Collection of Water 41

catchment. If these activities are suitably controlled they are unlikely to have any significant effect on the quality of water in

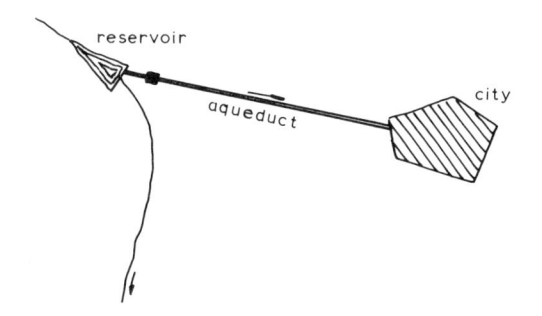

FIGURE 15. A direct supply reservoir

PLATE 6. Claerwen dam, a masonry structure belonging to the Birmingham Water Department

PLATE 7. One of the chain of masonry dams in the Elan Valley. There is a full width spillway

the reservoir but they create valuable improvements to the recreational facilities and amenities of the area. The direct-supply reservoir concept means that the water is conveyed to the demand centre in a closed aqueduct and is not therefore available for other potential uses. The aqueduct itself is usually an expensive structure because, by their very nature, upland catchments are often a considerable distance from centres of population. Thus, the water from the Elan Valley flows to Birmingham in an aqueduct which is 117 km long.

The river-regulation reservoir (Fig. 16) is an attempt to provide multi-purpose use of water from an upland catchment. The reservoir still stores excess flows in the river but, instead of the impounded water being conveyed in an aqueduct to the centre of demand, it is used to augment low flows in the river. When the flow at a control point downstream drops below some predetermined value, water is released from the reservoir to increase the natural flow and thus reduce the range of variation of the river flow. Storm flows which would otherwise escape to the sea are stored for discharge during dry weather thus improving the reliable yield of the river as well as reducing the incidence of flooding.

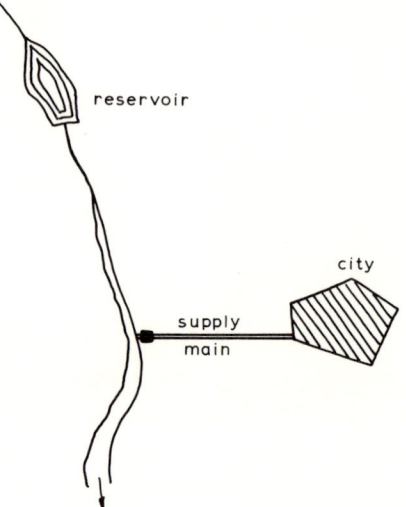

FIGURE 16. A river regulation reservoir

Collection of Water

Water is abstracted from the river at a point as close as possible to the demand and transport of the water is thus largely by means of the natural river channel so that expensive aqueducts are not required. Abstraction will have to be made by pumping from the river and further pumping is usually necessary to put the treated water into supply since the river will most likely be at a similar elevation to the city. With a direct-supply scheme the aqueduct is usually laid at a shallower gradient than the natural watercourse so that sufficient head is often available to obviate the need for pumping. This is an application on a larger scale, of the principle of the mill race which was used to gain head on a stream. As water flows down the river its quality deteriorates due to natural and man-made pollution so that more intensive treatment is necessary for the water obtained from a river regulation scheme than would be required for water drawn directly from an upland reservoir. However, since the water flows in the natural channel other users benefit from its presence and as full treatment is given to the water when it is abstracted there need be little or no restriction on amenity or recreational use of the reservoir. It will thus be appreciated that a river regulation scheme provides a more beneficial use of water resources for the community at large at the expense of increasing the cost of treatment necessary to produce a potable water.

The Clywedog scheme (Fig. 17) on one of the headwaters of the Severn is a good example of a river regulation reservoir used to impound storm flows which are then released, as necessary, to maintain the desired minimum flow at a gauging point downstream. Eleven water undertakings and the Central Electricity Generating Board benefit from the scheme and the reservoir is a popular tourist attraction with sailing and fishing facilities. A further benefit of the construction of the reservoir has been a significant reduction in flooding in towns downstream. It is of interest to note that the yield from Clywedog reservoir when used for river regulation purposes is about five times the yield which it could provide if used as a direct-supply reservoir, which highlights a further advantage of this type of development.

The older dams were usually built of masonry but most of the newer structures are of concrete (Plate 8), rock or earth depend-

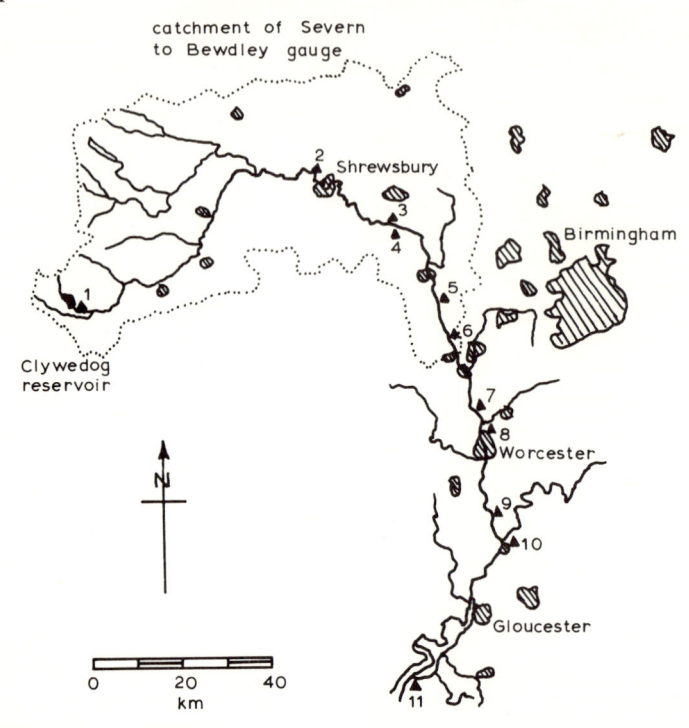

FIGURE 17. The Clywedog regulation scheme for the river Severn. This reservoir on one of the tributaries of the Severn is used to maintain a specified flow at Bewdley. Abstractors benefiting from the scheme are: 1 Montgomeryshire Water Board, 2 Shrewsbury Corporation, 3 East Shropshire Water Board, 4 CEGB Ironbridge Power Station, 5 South Staffordshire Waterworks Company and Wolverhampton Corporation, 6 Birmingham Corporation, 7 East Worcestershire Waterworks Company, 8 Worcester Corporation, 9 Coventry Corporation, 10 Cheltenham and Gloucester Joint Water Board, 11 Bristol Waterworks Company

ing on the site conditions and availability of materials. It may be argued that a large masonry or concrete dam can never happily blend into natural surroundings, and many of the modern earth dams with grassed downstream faces are less conspicuous (Plate 9). In any event, dams of all types and the reservoirs which they form are usually popular scenic attractions and may improve the economic situation in the neighbourhood by encouraging tourists. Most of the newer water supply reservoirs

PLATE 8. Clywedog dam of concrete construction which regulates the flow of the Severn. The spillway again covers the full width of the dam

PLATE 9. The earth dam at the Balderhead reservoir. The grassed downstream face is not particularly conspicuous. The spillway is a bellmouth type upstream of the dam face discharging into a tunnel through the dam which emerges in the centre of the picture. The upstream face of an earth dam is protected by an apron of concrete slabs or rocks

have thriving fishing and sailing clubs and participants in these activities may travel considerable distances to take advantage of the facilities. On at least one recently constructed reservoir special provision has been made for bird watching on part of the lake.

A problem met in both natural and artificial lakes which are deeper than 20 m is that of stratification which can be responsible for undesirable quality changes. In a stratified lake three zones are formed (Fig. 18).

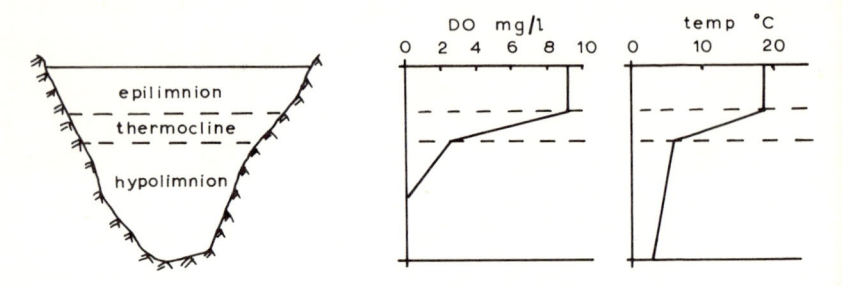

FIGURE 18. Stratification in a lake

(a) A circulation zone, with mixing due largely to wind action, in which there is good-quality water with high dissolved oxygen and a relatively high temperature. This zone is termed the *epilimnion*.

(b) A zone of rapid temperature drop, called the *thermocline*, which is usually only a few metres deep.

(c) A stagnant zone, the *hypolimnion*, where no mixing occurs. The temperature is low, dissolved oxygen is absent and the end products of anaerobic oxidation (hydrogen sulphide and methane) are present. As a result the water is of poor quality.

With the onset of cold weather in the autumn the surface water approaches the temperature at which it has a maximum density (3·94°C), thus creating an unstable situation. Wind action soon causes overturn so that the layers disappear and the lake is completely mixed with the good quality water becoming contaminated by the stagnant bottom water. Treatment problems are increased during this overturn period particularly due to the presence of tastes and odours in the water. In winter, stratification may occur due to the very cold surface water being less dense than the slightly warmer water below the surface. It is always desirable in an impounding reservoir to have draw-off facilities at several depths so that the best-quality water can be obtained at any particular time.

RIVER ABSTRACTION

The lowland reaches of large rivers are being increasingly used for water supply purposes. In these lower reaches the flow is much less variable than in the headwaters so that a relatively large safe yield may be available without the need for control works and, thus, the capital cost of abstraction is fairly small compared with impounding schemes. On the other hand, the water will have to be pumped into supply and treatment will be more extensive than for an upland supply so that operating costs tend to be higher. Because of the inevitable deterioration in quality down the river due to natural pollution from run-off as well as from effluent discharges, lowland river sources require careful scrutiny to maintain the desired water quality from the treatment plant. In flood time lowland rivers become highly turbid with turbidities in excess of 1000 mg/l due to the suspended soil and silt particles carried along in the flow and it may be necessary to stop abstraction during such periods. Storage of raw or treated water is thus necessary to provide continuity of supply in these circumstances.

A further potential hazard when dealing with lowland rivers is that accidental discharge of oil or toxic chemicals may occur requiring rapid action on the part of the authorities to prevent contaminated water from entering the supply. Usually, one of the first actions to be taken at a road accident involving the spillage of toxic or hazardous materials is for the fire brigade to hose down the area. Unfortunately, however, the contaminated run-off often finds its way into a ditch which may lead to a watercourse used as a raw water supply downstream. Because of these possible hazards and the need to cease abstraction during floods, many river abstraction schemes incorporate a storage reservoir, with a detention time of a week or two at normal demand rates, filled by pumping from the river. The storage capacity in the reservoir allows river abstraction to be stopped when dictated by flood conditions or when contamination of the river has occurred. In addition, the reservoir provides storage capacity for use in short periods of low flow and is often available for angling and sailing. These facilities are likely to be very popular because the reservoir will usually be relatively close to the urban area for which it supplies water.

In general, these storage reservoirs give a significant improvement in water quality due to settlement of suspended matter and die-off of bacteria because of the unsuitable environment. Sometimes, however, storage of a water containing nutrients such as nitrates and phosphates, which arise from agricultural run-off and sewage effluents can, in a shallow reservoir, result in prolific algal blooms. The water then becomes difficult to treat satisfactorily and tastes and odours are likely to be troublesome. The reservoirs may be constructed in small natural valleys or may be entirely artificial with enclosing earth embankments and need not be unattractive additions to the landscape.

GROUNDWATER

Certain natural strata provide reservoirs for the water which enters the ground by infiltration and it is possible to utilize these resources by collecting the discharge from springs or by drilling wells, which may be artesian, i.e., free flowing, or which may have to be pumped (Fig. 19). In England and Wales about 25 per cent of the water supplied for domestic and industrial use is drawn from groundwater sources, the two most important aquifers being chalk and Bunter sandstone. The water in these aquifers is replenished by infiltration and, thus, if water is abstracted at a rate greater than the rate of replenishment the water-table will fall and the source may be eventually exhausted. In the London area, where the chalk basin forms a large underground reservoir, over-abstraction has resulted in the need to pump from depths of over 100 m in wells which once flowed as artesian bores. In coastal areas, overpumping of aquifers can lead to the intrusion of seawater which contaminates the ground-

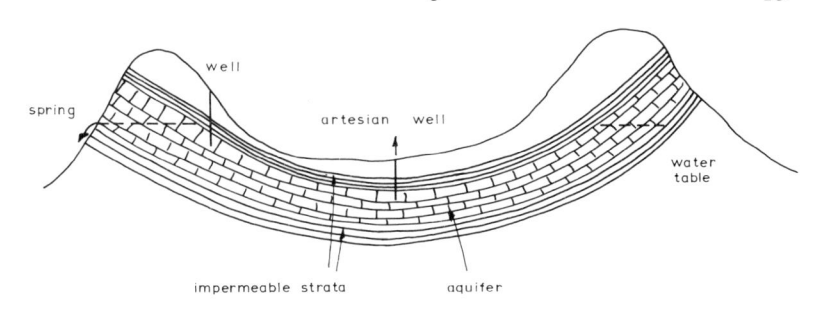

FIGURE 19. Groundwater abstraction

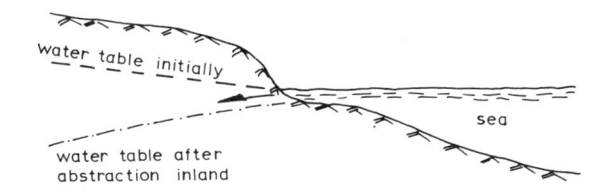

FIGURE 20. Intrusion of sea water into an aquifer due to overpumping near the coast

water and prohibits the further use of the aquifer for potable supply (Fig. 20). Artificial recharge of groundwater resources with river water or treated sewage effluents are possible remedies in such cases of over abstraction but care must be taken to ensure that suspended solids are removed from the recharge water as they may otherwise clog the voids in the strata. Conversely, it may be possible to abstract groundwater and use it to augment surfacewater flows as in the Lambourn Valley Scheme of the Thames Conservancy.

Groundwater is normally of high quality because of the efficient filtering action during the passage of the water through the strata although, in fissured strata, it is possible for polluted surfacewater to rapidly find its way to the water table. Groundwaters are usually hard due to the solution of calcium from the aquifer and they are often deficient in dissolved oxygen. Many contain iron in solution because of the absence of dissolved oxygen and, when exposed to the atmosphere, the iron is oxidized to an insoluble hydroxide giving a brown precipitate in the water. Iron in water at more than about 0·3 mg/l gives it a rather astringent taste and makes the brewing of tea most unsatisfactory. At even lower concentrations of iron, there is often trouble with staining of clothes in spin driers.

Assessment of the yield of groundwater resources is not easy because the extent and capacity of the aquifer are difficult to determine. Test pumping of exploratory boreholes is often used to predict the potential yield of an aquifer but interpretation of the results is difficult. The use of electrical analogues in which the flow of electricity through a network of resistors simulates groundwater flow is proving a useful technique for the solution of problems in this field. To enable such an electrical analogue (Plate 10) to be constructed it is necessary to know the hydraulic characteristics and the geology of the strata in some detail and collection of this information is not always easy.

When a borehole is sunk into a uniform aquifer and pumping is started, the water table is lowered around the bore producing a cone of depression (Fig. 21) which may increase the flow to the point of abstraction because of the steeper hydraulic gradient. Additional wells in the vicinity may interact with one another so that drilling a second bore adjacent to the first does not result in a doubling of the yield. Although the water table in permeable strata occurs at the surface as free water in lakes and rivers, abstraction of groundwater may so lower the water-table that the

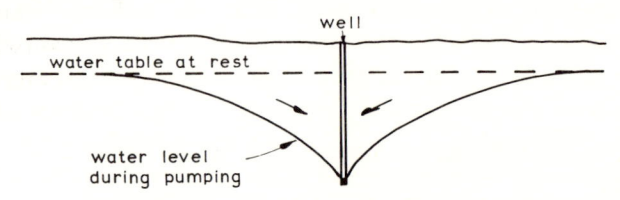

FIGURE 21. Cone of depression round a well

PLATE 10. A large electrical analogue model used to investigate three dimensional problems in groundwater flow

borehole is actually taking water from nearby watercourses (Fig. 22) and, even in natural conditions, there may be considerable interchange between surfacewater sources and groundwater. In many cases, the aquifer does not have uniform characteristics and the somewhat idealized concept of a cone of depression around a well is considerably modified.

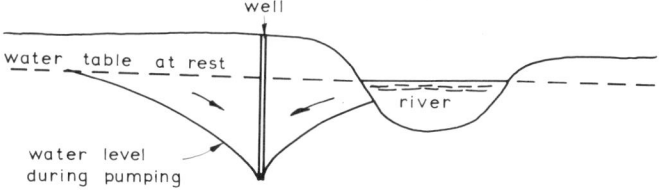

FIGURE 22. Induced re-charge of an aquifer from a river due to the effects of pumping on the water table

OTHER SOURCES OF WATER

Because of the shortage of conventional sources of water in some parts of the world attention is now being directed towards other possible sources, including sewage effluents and seawater, which because of their particular characteristics, require treatment methods rather different from those used for conventional raw waters. The possibilities and techniques of such water reclamation operations are discussed in Chapter 10.

5 Water Treatment

It will be clear from earlier chapters that water from all sources is likely to contain impurities which may make it unsafe to drink or, at least, make it aesthetically unpleasant. Although many industrial uses of water have fairly limited quality requirements some uses such as the manufacture of high-grade paper, food processing, etc., require water of potable or, in some respects, even higher than potable quality. It follows, therefore, that almost all raw waters require some form of treatment before use.

NATURE OF IMPURITIES

The types of impurities which are likely to be found in water are:

(*a*) Floating and large suspended solids, e.g., twigs, leaves and fish, etc.

(*b*) Small suspended and colloidal solids, e.g., clay and silt particles and micro-organisms.

(*c*) Dissolved solids, e.g., alkalinity, hardness, inorganic salts and organic compounds.

(*d*) Dissolved gases, e.g., hydrogen sulphide and carbon dioxide.

Figure 23 illustrates the range of sizes for some of these impurities and shows the approximate operational ranges of the main treatment processes. As will be noted, there is no individual process in the conventional treatment field which will remove all forms of impurity. It is thus necessary to suit the treatment process to the particular impurities to be removed. A combination of treatment processes is often necessary. Table 9 indicates the types of impurities likely to be found in water from various sources and shows the treatment processes required to produce an acceptable water. The most comprehensive treatment would be used for a lowland river water and Fig. 24 shows the flow diagram for such a plant which would have a capital cost of about £12/m³d and which would produce water at a cost of about 0·5p/m³.

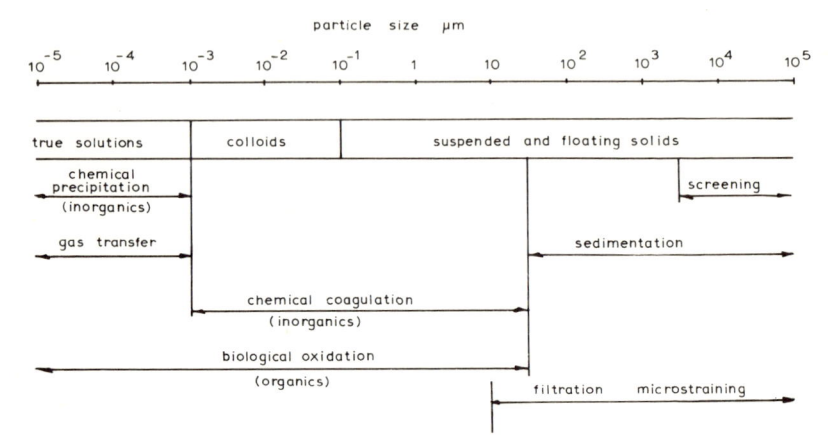

FIGURE 23. Particle sizes and effective ranges of treatment processes

Table 9. Probable treatment for water from various sources

Source	Probable impurities	Probable treatment
Upland catchment	colour	screening, microstraining or filtration, precautionary disinfection and colour removal by coagulation if necessary
Lowland river	micro-organisms, colour, turbidity and organic matter	screening, coagulation, sedimentation, filtration and disinfection
Deep well	hardness, absence of dissolved oxygen, iron and manganese	softening and precautionary disinfection, aeration and possibly catalyst filtration

Water Treatment

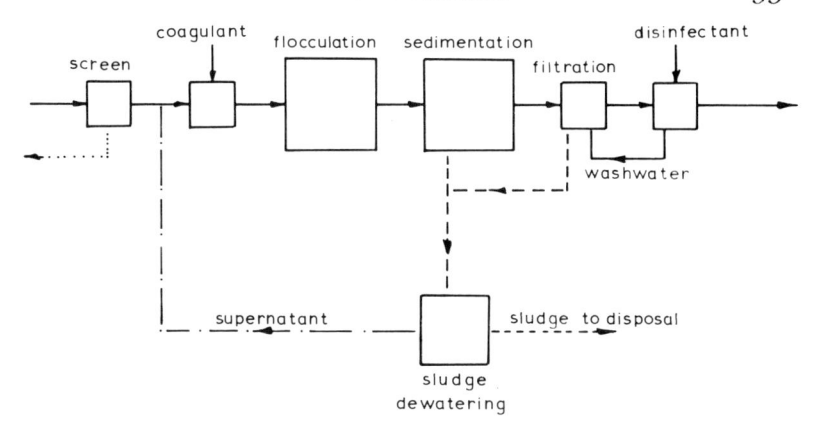

FIGURE 24. Flow diagram for a conventional water treatment plant

CONVENTIONAL WATER TREATMENT

PRELIMINARY TREATMENT

The intakes of a water treatment plant are usually protected by some form of floating boom and a coarse screen with openings of 25 to 50 mm which serve to exclude large floating and suspended objects like tree trunks and drift wood brought down by flood flows. The screening proper is carried out by passing the flow through a mesh of 5 to 20 mm aperture which may be arranged in the form of a drum, disc or continuous belt (Plate 11). The mesh removes leaves, paper and other trash which is washed off the screen as it is drawn up out of the flow, the trash being usually returned to the raw-water source downstream of the intake.

A more intensive form of screening is given by the microstrainer which uses a fine woven stainless steel mesh with apertures of 20 to 60 μm. The fabric is supported on coarser mesh around the periphery of a rotating drum (Plate 12) and because of its small aperture size is soon clogged. Continuous washing of the fabric is used to remove the screenings and requires about two per cent of the output of the strainer. Micro-strainers are able to remove the larger micro-organisms such as algae and may be capable of supplying all the treatment necessary, except

PLATE 11. Continuous band screens dealing with the raw water to a treatment plant. The straining mesh is inside the casing and is moved by the electric motor on the side of the casing

PLATE 12. Two microstrainers installed on a river-derived supply. The tops of the drums and the washing gear are visible above the floor, the remainder of the drums being submerged in water below floor level

Water Treatment

for disinfection, for waters normally of high quality but which occasionally contain large numbers of algae.

Storage of water for periods of 10 to 14 days usually has beneficial results on quality due to sedimentation of suspended matter and the death of many bacteria due to the lack of organic matter for food and the generally unsuitable environment. Algal blooms at certain times of the year in storage reservoirs can cause deterioration in the water quality and the addition of copper sulphate or other algicides may be a wise precaution to prevent heavy algal growths. Once large numbers of algae have appeared the problem cannot be immediately solved by killing the algae since they release taste- and odour-forming compounds as the cells disintegrate after death.

Table 10. Effect of particle size on settling velocity (particles of SG 2·65 in water at 10°C)

Particle diameter (μm)	Time to settle 1 m
1000	6 sec
100	3 min
10	3 hour
1	300 hour
0·1	1500 day
0·01	450 year

COAGULATION

Although some of the suspended matter present in raw waters would settle out if left to stand under quiescent conditions much of the material is colloidal and does not settle at all or only at negligible velocities (Table 10). To ensure satisfactory removal of turbidity due to colloidal silt particles, micro-organisms, etc., it is usually necessary to persuade these particles to coagulate into larger, heavier, and thus more readily settleable solids. By adding a suitable reagent which forms an insoluble precipitate it is possible to enmesh the colloidal particles in a spongy floc which, if gently stirred, flocculates into larger particles which have a reasonable settling velocity and can thus be removed by sedimentation. As an added benefit of coagulation, a certain amount of dissolved colour is also removed due to adsorption of the organic molecules on the surface of the floc particles. The usual coagulant for water treatment is aluminium sulphate (alum), $Al_2(SO_4)_3$ although iron salts are also used to some extent. The aluminium sulphate reacts with alkalinity in the water to give an insoluble hydroxide which forms the floc particles:

$$Al_2(SO_4)_3 + 3Ca(HCO_3)_2 = \underline{2Al(OH)_3} \downarrow + 3CaSO_4 + 6CO_2.$$

The operation of coagulation is carried out in two stages, addition of the coagulant solution with hydraulically or mechanically created turbulence to provide rapid mixing (Fig. 25), followed by flocculation achieved by gentle agitation with paddles or hydraulic turbulence (Fig. 26). The coagulant is fed either by metering pumps which deliver a saturated solution or by 'weigh batchers' which deliver dry powder to solution pots for injection into the flow. The appropriate coagulant dose for a particular water can only be found in laboratory experiments on a multiple stirrer jar-test apparatus (Plate 13) which simulates the rapid mixing, flocculation and sedimentation stages of the full-scale process. By adding different amounts of coagulant to samples of the water under test it is possible to determine the optimum dose of coagulant and the best pH for coagulation with that particular water (Fig. 27). In general, coagulation will only occur at the pH corresponding to maximum insolubility of the coagulant (about pH 7 for aluminium sulphate) and the higher the turbidity of the water the lower the coagulant dose required for coagulation.

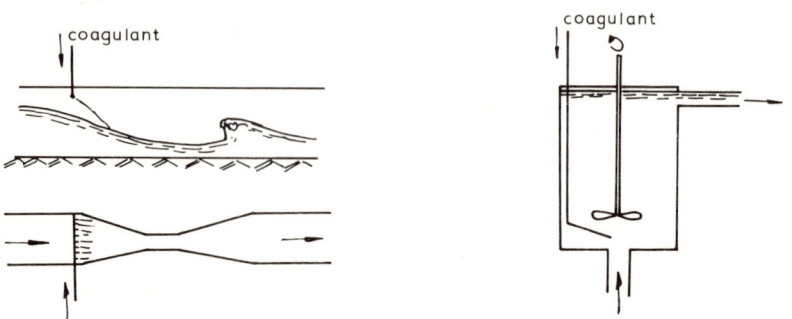

FIGURE 25. Rapid mixing. (a) By turbulence at a measuring flume, (b) By mechanical agitation

Water Treatment

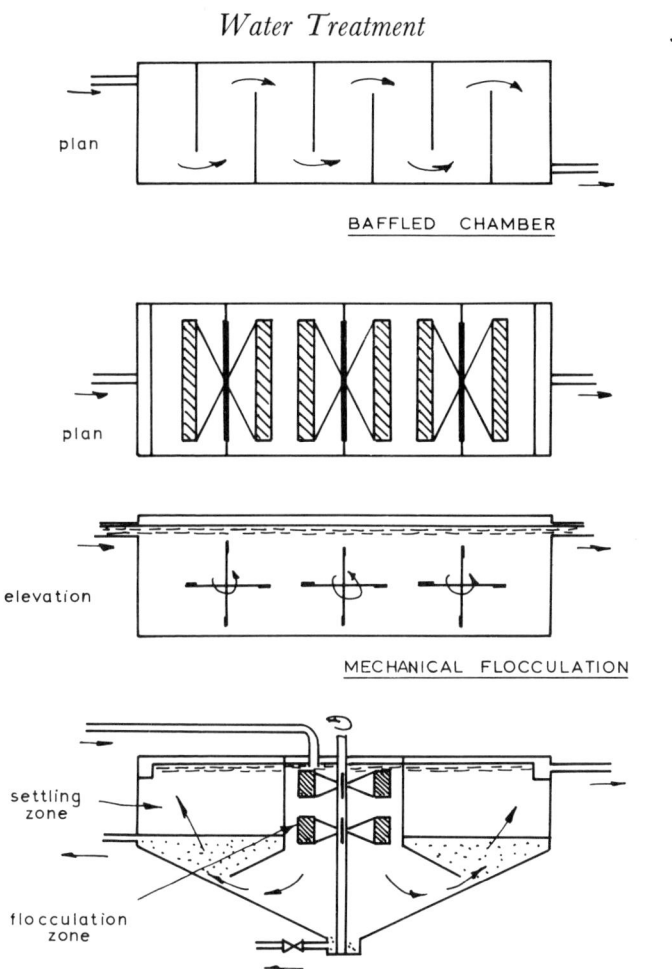

FIGURE 26. Methods of flocculation

With low turbidity waters difficulty is often experienced in achieving good coagulation and artificial turbidity in the form of bentonite may be added to provide nuclei for floc formation. Coagulant aids such as polyelectrolytes, which are long chain organic molecules, are sometimes used at low concentrations to improve floc formation or to reduce the coagulant dose required. Floc particles are relatively delicate so that water containing them must be conveyed to the sedimentation basins with care

PLATE 13. The jar-test apparatus for determining the chemical dose required for coagulation of water. The beakers in front contain water coagulated with increasing doses of aluminium sulphate from left to right, the clarity of the water in each beaker indicating the effectiveness of each particular dose

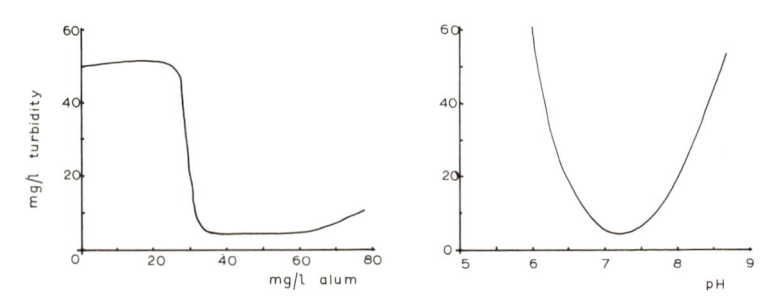

FIGURE 27. Typical jar-test results

since excessive turbulence will tend to shear the floc particles and result in a deterioration in their settling characteristics.

SEDIMENTATION

Suspended solids larger than a certain particle size, depending upon the material concerned, will have settling velocities which

Water Treatment

permit removal by sedimentation. When a particle is suspended in a fluid of lower density it will begin to accelerate under the action of gravity and will eventually reach a terminal velocity at which the gravitational force is exactly balanced by the frictional drag force on the particle as it falls through the fluid. For spherical discrete particles, i.e., particles which do not change in shape or other properties during settlement, in laminar flow conditions which normally apply in treatment plants, the terminal settling velocity v_s is given by Stoke's Law:

$$v_s = \frac{gd^2}{18\nu}\left(\frac{\rho_s - \rho_w}{\rho_w}\right),$$

where g = acceleration due to gravity;
 d = diameter of particle;
 ρ_s = mass density of particle;
 ρ_w = mass density of fluid;
 ν = kinematic viscosity of fluid.

In both water and wastewater treatment, flocculent particles are much more common than discrete particles and this complicates the analysis of settlement. Flocculent particles tend to agglomerate following collisions and thus tend to have increasing velocities with depth (Fig. 28). Under ideal conditions of uniform flow distribution and in the absence of short-circuiting, the limiting settling velocity of discrete particles removed in a sedimentation basin is given by the surface overflow rate

$$\left(= \frac{\text{rate of flow into basin}}{\text{surface area of basin}}\right).$$

Thus, for discrete particles, removal is independent of the depth of the tank. Because of the non-uniform settling characteristics of flocculent suspensions, greater removal of such suspensions occurs in deeper tanks since these offer more opportunity for contact between the particles. In practice, the hydraulic behaviour of settling tanks is far from the ideal plug-flow situation as can be demonstrated by flow-through curves (Fig. 29) obtained by adding a tracer to the inflow and plotting the measured tracer concentrations in the effluent from the tank. There is evidence however, that the hydraulic properties of the tank do not neces-

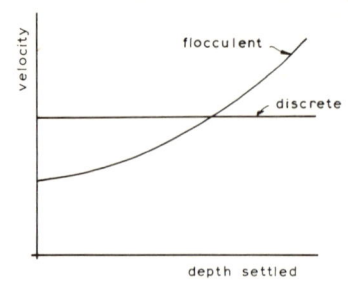

FIGURE 28. Settling behaviour of discrete and flocculent particles

deposited at least part of their solids load. The density currents may in some instances improve sedimentation since they can promote flocculation. In general, however, hydraulic turbulence due to inlet and outlet conditions as well as to density currents, together with the effects of wind on the surface mean that sedimentation tanks do not achieve their theoretical performance and allowance must be made for this in design.

sarily reflect its efficiency for the removal of settleable solids. Most of the short-circuiting which is responsible for the departure of the flow-through curves from the ideal plug-flow form is caused by density currents due to the differences in density between the inflow and the tank contents, which have already

Various forms of sedimentation tank are used for water treatment (Fig. 30), the vertical flow sludge blanket type (Plate 14) being particularly popular in the UK. Flocculation takes place in the lower portion of the tank and, as the particles rise, their upward velocity eventually becomes equal to their settling

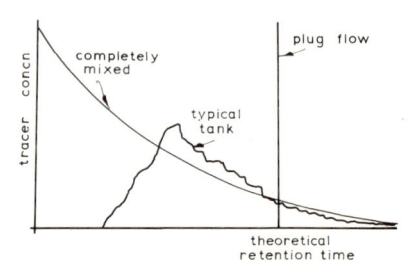

FIGURE 29. Flow-through curves for sedimentation tanks. The curve for a typical tank is a combination of the two extremes of plug flow and complete mixing, due to the effects of density currents and hydraulic turbulence

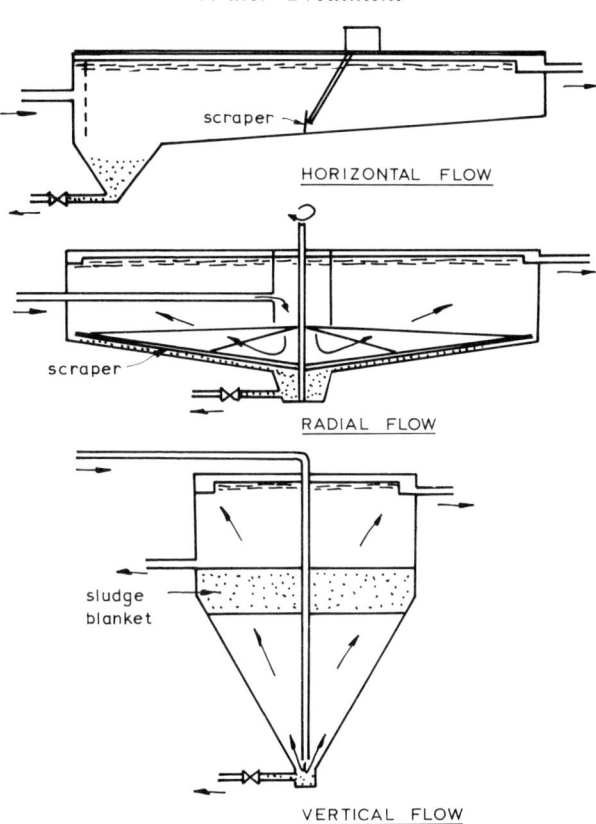

FIGURE 30. Types of sedimentation tank

velocity and they remain suspended at that level. A blanket of floc particles soon builds up and acts as a strainer to remove smaller particles which would otherwise escape. The clear water is removed from the surface by a system of channels arranged so as to promote even distribution of flow. Sludge is continuously removed from the blanket to prevent an excessive accumulation.

Because of its configuration, mechanical stirring or sludge removal equipment is not necessary and, thus, operation and maintenance of such tanks is simple. Their main disadvantage is that a hydraulic overload will increase velocities throughout the tank and may result in loss of the sludge blanket. Because of the

PLATE 14. A vertical flow sedimentation tank. Water with coagulant is fed into the bottom of the tank, coagulation occurs and the floc particles remain suspended in the tank while the clear water is removed by the decanting troughs across the surface of the tank

PLATE 15. A circular flocculation and sedimentation tank. Flocculation is induced by paddles within the chamber at the bottom left and the water then passes into the annular settling zone from whence the clarified water is decanted

steep slope (60°) of the sides the tanks are deep and construction may be expensive. American practice favours mechanical flocculation followed by sedimentation in horizontal or radial flow tanks of shallower depth than the vertical flow type. The units may be separate or may be combined in the same structure (Plate 15). The disadvantages of this solution are the increased complexity of the plant and the greater space requirements. Whichever method of sedimentation is employed the usual surface overflow rate for alum floc is about 30 $m^3/m^2 d$.

The sludge collected in sedimentation tanks must be removed carefully to avoid break-up of the floc and re-suspension of the solids. The denser the sludge which can be withdrawn the smaller the volume required for disposal which is a matter of considerable importance in present-day conditions. Sludge, and backwash water from filters, may simply be dumped in old quarries and similar declivities since it is of no value, but dumping facilities within reasonable distance of the treatment plant are becoming increasingly difficult to find and, in any case, their use for sludge disposal is not always desirable. Much more use is now being made, therefore, of the dewatering techniques already employed for dealing with sewage sludges and discussed in Chapter 9.

FILTRATION

It is often possible to remove about 90 per cent of the turbidity in a water by efficient coagulation and sedimentation. The remaining suspended matter is made up of colloidal solids and floc particles which were not removed by sedimentation and which would be undesirable in potable water partly from aesthetic reasons and partly because such particles might include harmful bacteria. To remove the remaining turbidity and give the water a final 'polish' it is passed through a bed of graded medium, usually 0·5 to 1·0 mm sand.

The earliest type of filter, the slow sand filter (Fig. 31) is loaded at about 2 $m^3/m^2 d$ and because of the low hydraulic loading, penetration of solids is superficial only. Removal of turbidity is enhanced by the formation of a microbial film on the surface which produces an efficient straining action and also causes biological oxidation of organic matter and nitrogen compounds which may be present in the raw water. With continued use the surface layers of sand become clogged so that water no longer flows through the bed at the desired rate. Thus, after a period of one to three months the filter is taken out of service and approximately 100 mm of sand is removed from the top, washed and replaced. Because of the biological activity in slow filters ammonia compounds are nitrified, thus simplifying disinfection of the filtered water. The large area of slow filters means that cleaning is a laborious process even though mechanical plant is available. Slow filters are thus unsuitable for waters with a high turbidity or for use after coagulation because floc 'carry-over' would rapidly cause clogging and more time would be spent cleaning the filter than in operation. The land area requirements are also a considerable disadvantage in these days of high land prices. Slow sand filters thus have only limited use today although some undertakings use them as secondary filters after preliminary treatment on the more heavily-loaded rapid sand filter described below. In this situation, slow filters with their biological action are able to oxidize undesirable organic compounds in the raw water and possibly prevent them causing tastes and odours. Indeed, the future may see the construction of more slow filters because their biological oxidation capability is an attractive feature when using polluted raw waters which have been subjected to preliminary treatment on rapid filters.

Rapid sand filters (Fig. 32, Plate 16) loaded at about 150 m^3/m^{2}d are the usual filtration units on modern plants. The relatively high hydraulic loading on these filters means that clogging occurs rapidly and the solids penetrate deeply into the

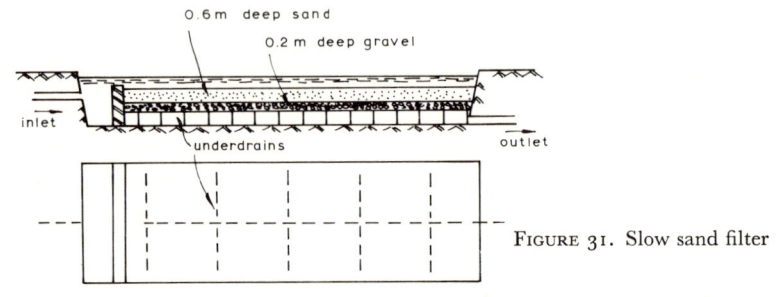

FIGURE 31. Slow sand filter

Water Treatment

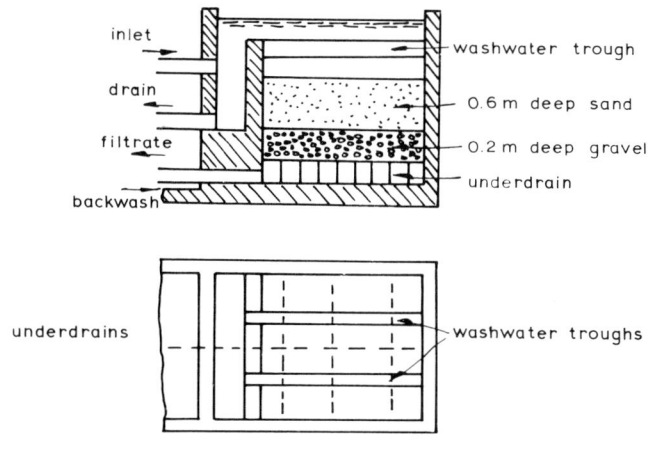

FIGURE 32. Rapid gravity sand filter

PLATE 16. A rapid gravity sand filter in operation. The sand is about 1 m below the water surface and the unit is controlled by the automatically operated valves in the upper-right

PLATE 17. Backwashing a rapid gravity filter. The dirty washwater is discharged to the central channel and removed for disposal

PLATE 18. Compressed air scour of a rapid gravity filter prior to backwashing. This technique helps to loosen material trapped in the bed and reduces the amount of water required for washing

Water Treatment

bed so that filter runs of 12 to 72 h before a head loss of about 2·5 m is reached are normal. Rapid filters are cleaned *in situ* by back-washing with filtered water at a rate of upward flow of about ten times the filtration rate (Plate 17). The upward force of the water suspends the sand grains as a fluidized bed and the voids between the grains increase in size. Trapped particles are then able to escape and the scouring action as the sand grains rub against one another ensures that all the solids are removed. Preliminary agitation of the bed with compressed air (Plate 18) helps to free trapped particles and reduces wash water consumption to about two to four per cent of the filter output. Because of the short runs obtained with rapid filters there is little biological activity and, thus, little if any nitrification takes place.

Most filters operate under gravity-flow conditions but in situations where pressure head needs to be conserved, e.g., in hilly areas (Fig. 33), pressure filters are used. These operate in the same manner as rapid gravity filters but are contained in a steel shell (Plate 19) which allows them to operate under heads of 100m or more. Coagulants, if required, are added to the filter inlet so that flocculation and coagulation occur in the sand bed. The media clogs more rapidly than in a gravity unit preceeded by sedimentation tanks and thus needs more frequent washing. The additional expense of more frequent washing is more than outweighed by the saving in pumping costs which would be incurred in hilly areas of supply if gravity filters were to be used.

Removal of suspended matter in filters is not simply a straining

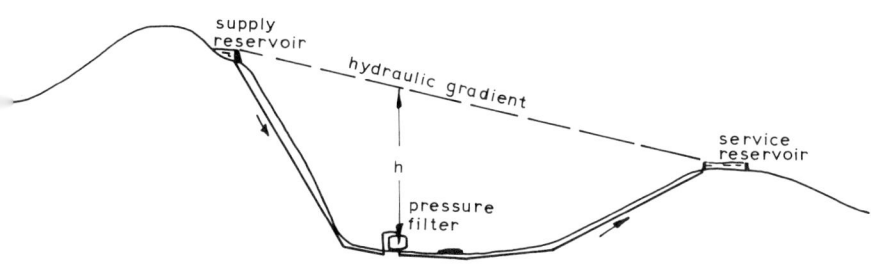

FIGURE 33. Rapid pressure filter application. Pressure filters can operate under head h thus allowing the service reservoir to be supplied by gravity. A gravity filter installation on the same site would require pumping to transfer water to the service reservoir

PLATE 19. A rapid pressure filter. The sand is contained within a steel shell capable of withstanding considerable pressure so that the filter can operate under a large hydraulic head

process, since solids much smaller than the voids in the bed are removed. Removal of solids is, in fact, due to a combination of adsorption, sedimentation and molecular attraction. The head loss characteristics in porous media filters are complex, since at any particular time the overall head loss across the bed is the sum of the head loss across the clean bed plus an additional head loss due to the deposited solids in the bed (which reduce the voids in the bed and, hence, increase the head loss). These two factors result in an exponential relationship between head loss and the volume filtered (Fig. 34). Because of this time variant head loss characteristic, automatic control gear is usually fitted to filters to ensure constant output without the need for continual manual adjustments to discharge valves.

When a graded filter medium is backwashed the effect is to produce a stratified bed since on cessation of backwash flow the coarse particles settle at a higher velocity than the fine particles and, thus, the bed is graded from fine particles at the top to coarse particles at the bottom. This is clearly not an ideal situation for filtration since the smallest voids, which have the highest head

Water Treatment

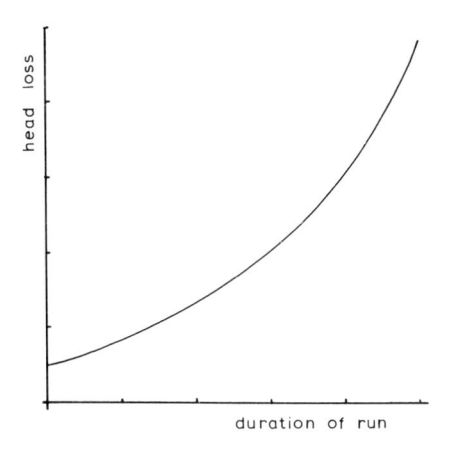

FIGURE 34. Filter head loss curve. The exponential increase in head loss is caused by the progressive clogging of the voids in the bed and the build up of surface deposits

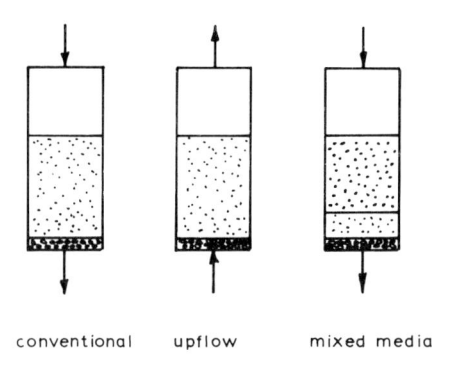

FIGURE 35. Modified forms of filter. In a conventional filter water meets the finest particles first causing rapid build up of head. The upflow filter operates in a more rational manner by passing water through the large particles first and reserving the fine voids for particles which are not removed in the lower parts of the bed. The mixed media filter employs large low density particles with low head loss to provide the initial removal. The finer and denser particles below give a final polish to the water

loss, are on the top of the filter and are thus used to trap both fine and coarse suspended solids. As a result the filter becomes unable to maintain flow long before the full void capacity of the bed has been utilized. In a rational filtration system the suspension would pass through the coarse voids first and only the particles which were fine enough to pass through any one layer would proceed to the next layer of smaller voids.

Developments in filter technology aimed at producing more efficient use of the bed without reducing the quality of the filtrate are being adopted in many new treatment plants (Fig. 35). Upward flow filtration gives the desired void configuration for efficient filtration but care must be taken to control filtration rates so that the upward flow does not expand the bed in the same way as does backwashing since this would result in poor filtrate quality. Grids in the top of the bed can be used to hold the sand in place during filtration but allow it to expand during backwashing. Multi-layer beds operate in the conventional down flow manner but are composed of two or more media with different characteristics, e.g., 0·5 to 1·0 mm sand (SG 2·6) and 1·0 to 2·0 mm anthracite (SG 1·6). The anthracite, although larger than the sand, has a much lower density and remains on top of the sand after backwashing. The large voids in the anthracite produce little head loss compared with the sand (Fig. 36) but provide a coarse filtration to remove the large suspended solids which would otherwise quickly block the sand. The filter run is thus extended for the same rate of flow or, more important on overloaded works, the rate of flow can be increased without reducing the length of filter run.

The length of filter runs with any type of unit may be determined by one or more of three criteria.

(*a*) *Terminal head loss*, usually about 2·5 m. When the filter reaches this head loss the run is stopped. (*b*) *Filtrate quality*. Breakthrough of suspended solids may occur when the bed is clogged and is indicated by an increase in filtrate turbidity possibly before the terminal head loss has been reached. (*c*) *Duration of run*. This may be predetermined on the basis of operating experience at, say, 24 or 48 hours.

Water Treatment

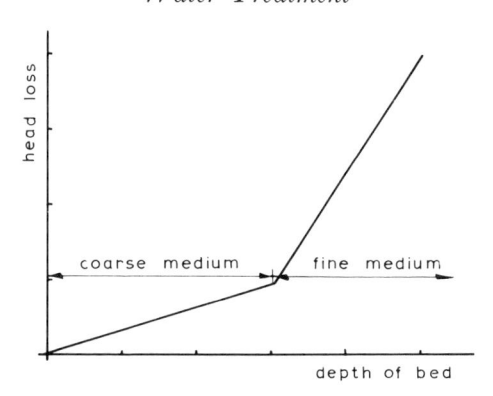

FIGURE 36. Head loss characteristics of a mixed media bed

The first two methods are suitable for plants where the raw-water quality is variable; whereas the time method is really only suited to raw waters of reasonably constant turbidity. Whichever method is adopted filter operation can be entirely automatic.

DISINFECTION

Although many micro-organisms will be removed by coagulation and sedimentation some will escape enmeshment in floc particles and because of their small size, may pass through a sand filter. In order to ensure the safety of the supply it is essential to destroy harmful micro-organisms by a disinfecting process. It should be noted that sterilization, which implies the destruction of all organisms, is not normally achieved or, indeed, required in water treatment.

The desired degree of disinfection is based on the concept of the percentage kill of micro-organisms required to produce a treated water containing not more than the specified number of micro-organisms.

The usual disinfecting agent for water supplies is chlorine or one of its compounds since they are cheap and effective disinfectants which are harmless to man at the concentrations used in drinking-water treatment (usually 1 to 2 mg/l) but are highly toxic to bacteria. Chlorine may be obtained as a gas, liquid chlorine under pressure or as a solution of sodium hypochlorite (bleaching powder), the latter method being particularly useful in emergencies or for small supplies. It is always converted to a chlorine solution before application to the water. Since chlorine is a powerful oxidizing agent it will react with many organic compounds or reducing agents which may be present in the water, e.g.,

$$4Cl_2 + H_2S + 4H_2O = H_2SO_4 + 8HCl.$$

This chlorine demand must be largely satisfied before a chlorine residual remains for disinfection. In the absence of ammonia, a free chlorine residual is then formed:

$$Cl_2 + H_2O \rightleftharpoons HCl + HClO$$
$$\updownarrow \qquad \updownarrow$$
$$H^+ + Cl^- \quad H^+ + ClO^-,$$

hypochlorous acid HClO being the more efficient disinfectant and predominating at acid pH values. In the presence of ammonia the actions are more complicated and combined residuals are formed:

$$Cl_2 + NH_3 = NH_2Cl + HCl$$
monochloramine

$$NH_2Cl + Cl_2 = NHCl_2 + HCl$$
dichloramine

$$NHCl_2 + Cl_2 = NCl_3 + HCl$$
nitrogen trichloride.

Combined residuals are not as potent as free residuals but are more stable and, thus, remain in solution longer. To give the same degree of disinfection, with the same contact time, a combined residual must be about 25 times the concentration of a free residual. Combined residuals are sometimes useful however, because they are less likely to produce tastes and odours after chlorination since they do not combine with phenols. Ammonia is sometimes added to waters lacking in ammonia so that combined residuals will be produced. Such residuals do not, however, remove tastes and odours already present. Superchlorination with comparatively large doses of chlorine (up to 20 mg/l) can oxidize objectionable taste- and odour-producing compounds after which the excess chlorine is removed to the desired residual by dechlorination, usually with sulphur dioxide:

$$Cl_2 + SO_2 + 2H_2O = H_2SO_4 + 2HCl.$$

Water Treatment

Superchlorination is also useful with waters of variable quality, since sufficient chlorine can always be added to ensure satisfactory disinfection, the excess being removed as necessary using an automatic chlorine residual controller.

If further chlorine is added to the water containing combined residuals these are oxidized so that the chlorine residual drops but then increases again, this time in the form of a free residual. This action produces the breakpoint curve (Fig. 37). For all practical purposes the breakpoint, i.e., the point of minimum residual, occurs at a chlorine dose of about 10 times the initial concentration of ammonia in the sample.

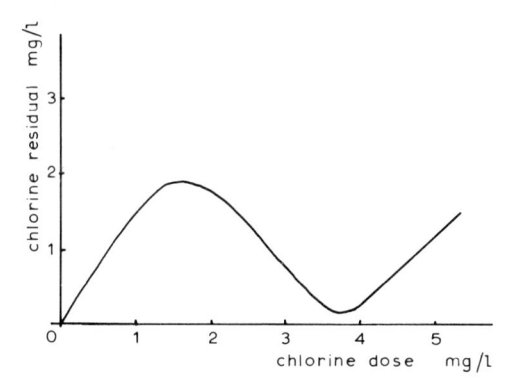

FIGURE 37. The chlorine breakpoint curve. The water contains 0.5 mg/l ammonia and the initial residual is of the combined form. Further addition of chlorine past the breakpoint produces a free residual

Because of the dangerous nature of chlorine gas and the corrosive nature of chlorine solutions, care is necessary in their handling. Chlorine is normally metered as a gas under vacuum conditions and then dissolved in a small volume of water which is injected into the supply (Plate 20). For small supplies the chlorine is used from gas cylinders but for large installations liquid chlorine is delivered under pressure in drums the contents being allowed to evaporate to provide gaseous chlorine. Bulk tanker deliveries of liquid chlorine are used at some new water-treatment plants.

PLATE 20. A modern chlorinator room. The chlorine gas is measured by visible flowrate meters in the centre of each panel and the residual chlorine in the water after treatment is shown on the circular dials. The room is sealed off from other parts of the building and provided with an efficient ventilation system to reduce the hazard from leaking chlorine gas

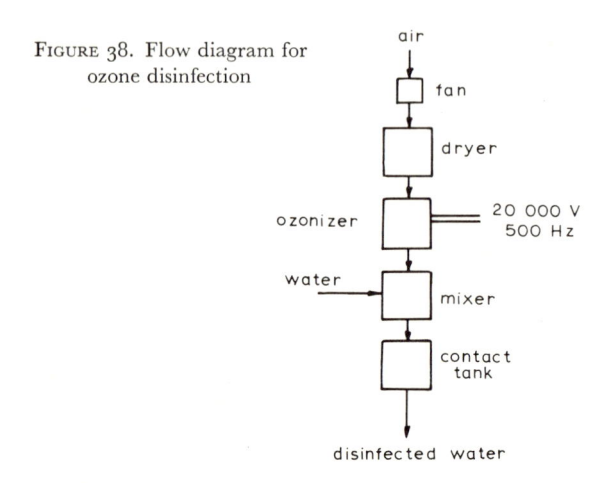

FIGURE 38. Flow diagram for ozone disinfection

Water Treatment

The only other disinfectant used in significant quantities for potable water supplies is ozone (O_3), an allotrophic form of oxygen produced by electrical discharges (Fig. 38). Ozone is a very reactive gas, a powerful oxidizing agent and an efficient disinfectant although, due to its reactivity, it leaves no residual in the water. In most situations ozone is considerably more expensive than chlorine and its use tends to be limited to isolated sites where transport costs would make chlorine expensive or to areas where cheap hydro-electric power is available. Some ozone installations have been used on highly coloured upland catchment supplies where colour removal by the oxidizing action of ozone is an important side effect of ozonization.

In order to produce sterile water for certain industrial purposes, e.g., the manufacture of pharmaceuticals, water may be sterilized by heat or by ultra-violet radiation. The cost of such methods make their use for conventional supplies quite uneconomic apart from the fact that sterilization by heat has a detrimental effect on the palatability of the treated water.

SOFTENING

For domestic water supplies, softening may be carried out as an economic measure since it cannot be justified on grounds of potability. Indeed, a hard water sometimes has a better taste than a soft water. Hardness which is due mainly to calcium and magnesium salts is often a problem with groundwaters where the hardness may exceed 300 mg/l. Regardless of its actual form, hardness is always expressed as calcium carbonate as a convenient common denominator. Removal of hardness can be achieved by precipitation methods or by the use of ion-exchange techniques.

Precipitation softening converts the soluble hardness into insoluble compounds which can be flocculated and removed as a sludge in sedimentation tanks. The reagents necessary for precipitation softening depend on the form in which the hardness is present, e.g., for calcium bicarbonate hardness lime is used:

$$Ca(HCO_3)_2 + Ca(OH)_2 = \underline{2CaCO_3 \downarrow} + 2H_2O.$$

Calcium carbonate precipitates out leaving a residual of about

40 mg/l due to its solubility at normal temperatures and the limited contact time available in a continuous flow plant. The softened water is saturated with calcium carbonate and, to prevent deposition of carbonate scale in pipelines, the water is stabilized by carbonation with carbon dioxide or by the addition of phosphates. The steps in lime softening are shown in Fig. 39. When calcium is combined with radicals other than bicarbonate, carbonate in the form of soda ash (Na_2CO_3) is provided to allow precipitation of calcium carbonate in the lime-soda process:

$$CaSO_4 + Na_2CO_3 = \underline{CaCO_3} \downarrow + Na_2SO_4.$$

For magnesium hardness of the carbonate variety, excess lime must be added to give a pH of about 11 at which value magnesium hydroxide precipitates leaving a residual of about 10 mg/l (magnesium carbonate is soluble in contrast to calcium carbonate):

$$Mg(HCO_3)_2 + Ca(OH)_2 = MgCO_3 + \underline{CaCO_3} \downarrow + 2H_2O;$$

at pH 11

$$MgCO_3 + Ca(OH)_2 = \underline{Mg(OH)_2} \downarrow + \underline{CaCO_3} \downarrow.$$

The high pH required for the production of magnesium hydroxide also ensures quite efficient disinfection since microorganisms are generally unable to live under such alkaline conditions. The excess lime-soda process must be used if hardness due to both carbonate and non-carbonate magnesium compounds is present. Large volumes of sludge are produced from precipitation softening processes and may pose disposal problems although calcium sludges may have some value for agricultural purposes. Alternatively, it is possible to recover lime from the sludge by heating although the cost of this reclaimed lime may exceed the cost of lime from normal commercial supplies.

Ion-exchange softening utilizes the property of certain natural materials such as greensands and zeolites, and of more efficient synthetic compounds, to exchange ions in their structure for different ions to which they are exposed. Natural zeolites are complex sodium-alumino silicates and may be represented as Na_2X where X indicates the complex structure. When hard water

Water Treatment

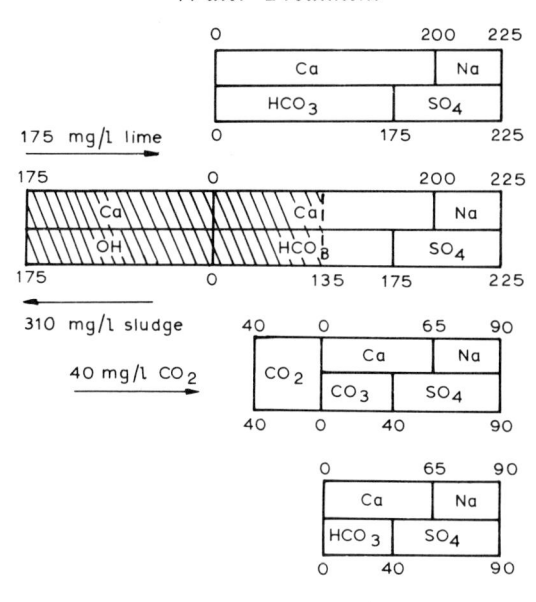

[Figure showing lime softening bar diagrams with values labeled in mg/l CaCO₃]

all in terms of mg/l CaCO₃

FIGURE 39. Lime softening. Hardness in this example is reduced from 200 mg/l to 65 mg/l. It will be noted that noncarbonate hardness is not removed by lime softening

is placed in contact with the zeolite, sodium is exchanged for the calcium or magnesium in the water and a water of zero hardness results:

$$Ca(HCO_3)_2 + Na_2X = CaX + 2NaHCO_3.$$

Eventually, the exchange capacity of the zeolite becomes exhausted and no further softening takes place. The ion-exchange material can, however, be regenerated almost indefinitely by using a strong salt solution to give an excess of sodium ions which displace the calcium and magnesium ions from the structure:

$$CaX + 2NaCl = Na_2X + CaCl_2.$$

The exchanger is then ready for further use, but it should be noted that a waste solution of calcium (or magnesium) chloride is produced, and suitable arrangements must be made for its disposal. The ion-exchange material is usually installed as a bed in a metal shell similar to a pressure filter.

The ion-exchange reaction described above is known as a sodium-cycle cation exchange, i.e., sodium ions are exchanged for other positively charged ions. A hydrogen cycle cation exchanger, which may be a natural or synthetic carbonaceous resin, is used to remove all cations except hydrogen ions from water as a preliminary to the production of demineralized water:

$$\begin{matrix} Ca \\ Mg \\ Na \end{matrix} \begin{matrix} CO_3 \\ SO_4 \\ Cl \end{matrix} + H_2Z = \begin{matrix} Ca \\ Mg \\ Na \end{matrix} Z + \begin{matrix} H_2CO_3 \\ H_2SO_4 \\ HCl. \end{matrix}$$

Regeneration is by acid treatment:

$$\begin{matrix} Ca \\ Mg \\ Na \end{matrix} Z + H_2SO_4 = H_2Z + \begin{matrix} CaSO_4 \\ MgSO_4 \\ Na_2SO_4. \end{matrix}$$

Since the water now contains only acid radicals, an anion exchange resin, i.e., a resin which will exchange negatively charged ions, usually a synthetic ammonia based organic compound, RNH_3 where R represents the organic part of the molecule, can be used to give demineralized water:

$$\begin{matrix} HNO_3 \\ H_2SO_4 \\ HCl \\ H_2SiO_3 \\ H_2CO_3 \end{matrix} + RNH_3^+.OH^- = RNH_3 \begin{matrix} NO_3 \\ SO_4 \\ Cl + H_2O \\ SiO_3 \\ CO_3. \end{matrix}$$

Regeneration is achieved by using a strong alkali:

$$RNH_3 \begin{matrix} NO_3 \\ SO_4 \\ Cl \\ SiO_3 \\ CO_3 \end{matrix} + NaOH = RNH_3^+.OH^- + \begin{matrix} NaNO_3 \\ Na_2SO_4 \\ NaCl \\ Na_2SiO_3 \\ Na_2CO_3. \end{matrix}$$

Such water is necessary for the feed-water to high-pressure boilers where any impurities can cause scaling and corrosion. Small demineralizing plants are frequently used to produce so-called distilled water for laboratory use. Cation and anion resins may be combined in a single bed to give demineralization in one unit. Although ion-exchange treatment does not produce sludge it does give rise to waste liquid streams after regeneration and, whilst these are often accepted into public sewers, they could be troublesome in isolated areas.

SPECIALIZED TREATMENT PROCESSES

The treatment processes described above are those normally employed in water treatment but in particular circumstances certain other processes may be necessary.

REMOVAL OF TASTES AND ODOURS

For potable supplies and water used in the manufacture or processing of food and drinks, tastes and odours due to dissolved gases and organic compounds may be troublesome. The contaminants may be due to man-made pollution, e.g., phenols from tar spraying operations which combine with chlorine to form chlorophenols with very strong medicinal tastes. On the other hand, natural sources such as algae growing in reservoirs can cause many tastes and odours in water. Prevention is usually better than cure in the case of tastes and odours so that the ideal solution is to prevent the discharge of potentially troublesome wastes and to control growths of algae by dosing reservoirs with copper sulphate or other herbicides before algal blooms appear.

In the event of tastes and odours being present in the water they can often be removed by the use of powerful oxidizing agents such as chlorine, chlorine dioxide or ozone. In some cases, adsorption of the offending material by activated carbon may be more suitable. Activated carbon in powder form may be added to flocculation basins or to filters in doses of 10 to 50 mg/l and will efficiently remove most tastes and odours although the cost of activated carbon is high and it is lost in the sludge or backwash water. In extreme cases it may be necessary to use beds of activated carbon which can, however, be re-used after regeneration by heating to about 1000°C, thus burning off the adsorbed organic material.

IRON AND MANGANESE REMOVAL

Groundwaters often contain iron and manganese in solution and impounded waters may contain soluble manganese if stratification has occurred because the absence of dissolved oxygen in the hypolimnion promotes the solution of manganese from organic deposits. Iron is soluble only in the absence of dissolved oxygen and at pH levels below 6·5. Neither element produces a health hazard at the levels normally found in water but their presence is responsible for taste problems and in even lower concentrations (about 0·2 mg/l) can cause staining of clothes in washing machines and spin driers, which act as highly efficient filters to remove colloidal solids from water with unfortunate results for the cleanliness of the washing.

Iron is usually removed by aeration of the water and pH adjustment to above 6·5. This causes oxidation of the iron and its precipitation as ferric oxide which may be removed by sedimentation or filtration. Manganese can also be removed by oxidation although more intense aeration is required and filtration through a catalyst medium, e.g., pyrolusite containing manganese, is often necessary with adjustment of pH to 6–7. Chlorination may be used instead of aeration prior to catalyst filtration.

FLUORIDATION

Evidence from various parts of the world suggests that the presence of fluoride in water has a retarding effect on dental decay particularly for young children. The level of fluoride considered to be beneficial is 1 mg/l since higher levels can produce mottling of the teeth. At high concentrations of fluoride (about 50 mg/l) severe bone damage can occur. Fluorosis is an occupational disease in aluminium smelting where fluorides are used as fluxes. Some natural waters contain fluoride and it has been recommended by the Ministry of Health that supplies deficient in fluoride should be artificially fluoridated to 1 mg/l and that supplies containing more than 1 mg/l should be reduced to that level. It should be noted that the recommendation has aroused considerable argument in both scientific and lay quarters not only on the question of whether compulsory mass-medication is desirable but also because of doubts over the long-term effects of consuming artificially fluoridated water. There is some concern

Water Treatment

that the artificially added fluorides may not behave in the body in the same manner as the naturally occurring calcium fluoride. Unfortunately, calcium fluoride has a very low solubility and its use is thus not practicable for fluoridation.

Fluoridation is usually achieved by injecting a fluoride solution made from sodium silicofluoride which is the cheapest source of fluoride available. The low solubility of sodium silicofluoride necessitates very accurate dry-feeding equipment and for small flows the more expensive sodium fluoride may be preferable because of its higher solubility, 40 g/l as against 4 g/l. Whichever form of addition is adopted, careful control and monitoring of the process is essential to maintain the desired fluoride level of 1 mg/l.

CORROSION CONTROL

A water capable of depositing a thin film of calcium carbonate is normally considered as non-corrosive since the film prevents corrosion of metal pipelines and fittings. The stability of a particular water with respect to the deposition of calcium carbonate scale is denoted by the Langelier Saturation Index:

$$LSI = \text{actual pH of water} - pH_s$$

where pH_s = pH which the water would have if it were saturated with calcium carbonate.

A positive saturation index means that deposition of scale is likely and, thus, the water will tend to be non-corrosive. Conversely, a negative index would indicate an unstable corrosive water. Waters softened by precipitation methods are usually non-corrosive but ion-exchange softened waters are often corrosive. Stabilization is achieved by removing excess carbon dioxide by aeration and adding sodium carbonate to produce a pH of about 7·8.

Very soft upland catchment waters may be highly acidic and in the days of lead pipes, plumbosolvency was a problem. The corrosive nature of such waters is countered by the addition of lime and sodium carbonate to give an alkaline pH. Fittings made of duplex brass, i.e., those containing 35 to 45 per cent zinc, are often subject to rapid corrosion by dezincification which leaves a porous fitting. This material is now obsolescent for fittings but, unfortunately, many are still in use. In the next chapter consideration is given to the subject of corrosion control in pipelines.

6 Distribution of Water

After water has been treated it is necessary to distribute it to all those served by the area water undertaking. The construction and maintenance of a water distribution system for a large city is a costly and complex operation since there will be at least one water main in each street. Many of these mains may be old and likely to fail if the surrounding ground is disturbed by excavations for new services or by the vibrations caused by the ever increasing flow of traffic. The basic elements of a typical distribution system are shown in Fig. 40.

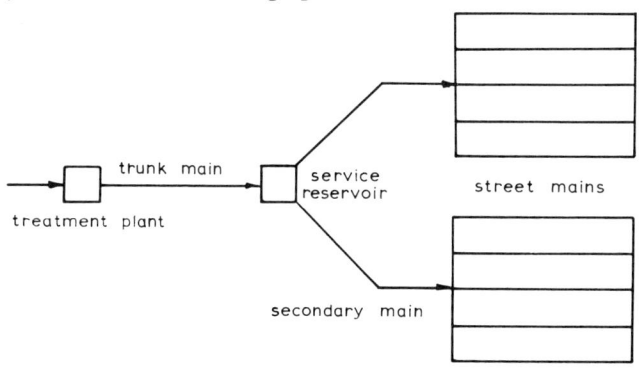

FIGURE 40. Elements of a water distribution system

WATER MAINS

Water mains can be divided into three classes;

(a) *Trunk mains*, the main supply lines between the treatment plant and service reservoirs or water towers. They are not normally used for direct consumer connections.

(b) *Secondary mains*, which distribute water from the service reservoirs to the street service mains and which in some cases provide bulk supplies to large industrial consumers.

(c) *Service mains*, the pipes along each street to which the individual consumer connections are made.

All water mains flow under pressure and the classical formula for the loss of head in such pipes is:

$$H_l = \frac{flv^2}{2gd},$$

where H_l = head loss in length l;

f = friction factor depending on pipe age and material;
v = velocity of flow;
d = diameter of pipe.

This formula illustrates the importance of selecting the appropriate pipe diameter for a particular situation. For a given flow, reducing the pipe diameter by half results in a 32 fold increase in head loss.

Water mains remain in use for many years so that roughening of the surface due to corrosion is likely and this will increase the frictional head loss. This, and also the effect of numerous joints between pipes on the friction characteristics, necessitates employing an empirical pipe flow formula. The Hazen-Williams formula is a widely-used empirical formula developed for water mains:

$$Q = 0 \cdot 91 C d^{2 \cdot 63} s^{0 \cdot 54},$$

where Q = flow in m³/s;
C = a constant depending on pipe roughness;
d = pipe diameter in m;
s = loss of head per unit length of pipe.

It will be noted that, for a given set of conditions, the flow is proportional to the value of C. Typical values of C for various types of pipe are given in Table 11 and show the marked reduction in capacity of old cast-iron pipes which have become roughened by corrosion. Various types of pipe are used for water mains, including.

(a) *Cast iron*. Most old mains were laid in cast-iron and many have given good service for over 100 years. Its main disadvantages are brittle nature, high weight, availability only in short lengths so that jointing costs are high and rough internal surface which gives a relatively high head loss. Cast-iron pipes are no longer produced but many fittings are still made in cast-iron.

Table 11. Typical pipe roughness factors

Pipe	Hazen-Williams C (SI units)
Coated cast-iron:	
new	75
30 years old	55
100 years old	40
Coated mild steel, new	75
Asbestos cement	80
Concrete	75
Plastic	80

(*b*) *Spun iron.* Centrifugal casting techniques produce lighter pipes with greater strength and smoother surfaces. The use is increasing of ductile cast-iron pipes which are much less brittle than ordinary cast-iron and which can withstand considerable tensile stresses without failing.

(*c*) *Steel.* Welded sections may be used to give pipes of almost any length which are relatively light and flexible. They have a smooth surface and high tensile strength. The major disadvantage of steel pipes is that the thin section can be rapidly damaged by corrosion if efficient preventive measures are not taken.

(*d*) *Asbestos cement.* These pipes are formed by applying pressure to a mixture of asbestos and Portland cement in suitable moulds. The resultant pipes are dense and smooth with high corrosion resistance. Asbestos cement is, however, rather brittle and can easily be broken by careless backfilling of trenches.

(*e*) *Concrete.* Usually made by spinning concrete in a mould to give a dense watertight pipe. May be reinforced with a steel cage or prestressed by wrapping with high tensile steel wire under load. Connections are difficult to make to concrete pipes because, particularly in the case of prestressed pipes, cutting will damage the reinforcement.

(*f*) *Plastics.* Pipes of up to 1200 mm diameter are now available in plastics which give light, smooth, flexible pipes, easily jointed by special solvents. The extra cost of plastic pipes over more conventional materials is counterbalanced by the ease with which they can be handled and laid. One man can lift a length of this pipe which, if made of metal or concrete, would require a gang of men or a mobile crane. Plastics pipes are virtually impervious to corrosion by the normal constituents of water and soil.

Except for welded steel mains, pipes are usually laid in short lengths of 3 to 6 m and joints inserted between each length (Fig. 41). These joints may be rigid bolted flanges (normally used for connecting valves and other 'specials'), semi-rigid spigot and socket joints, or flexible joints with rubber or plastic sealing rings which allow a certain amount of movement between adjacent pipes. Flexible joints allow many curves to be negotiated without the use of special bends and they are essential in ground conditions where subsidence is possible.

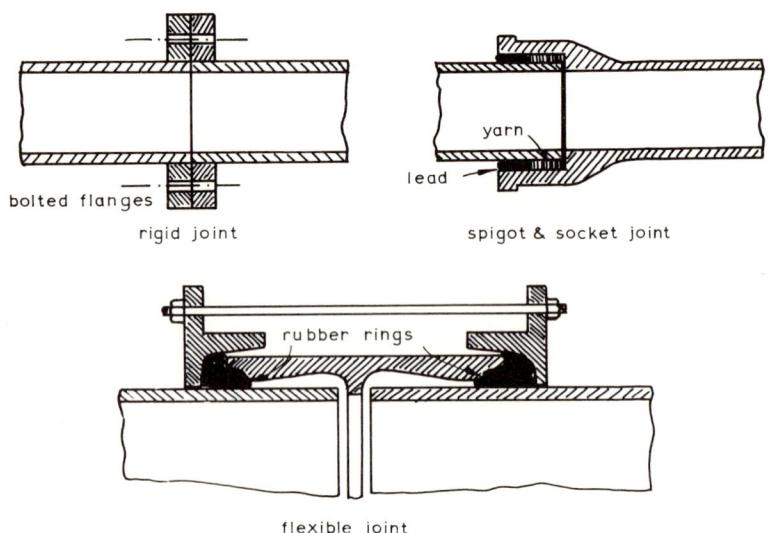

FIGURE 41. Pipe joints for water mains

In any particular situation there will be a range of pipe diameters which would carry the desired flow and when a pumping main is to be built careful design is required to select the optimum size of pipe. As shown in Fig. 42, the smaller the pipe

Distribution of Water

diameter the higher the head loss and, hence, the higher the cost of pumping. A smaller pipe however, has a lower capital cost than a large pipe although the latter would reduce pumping costs. All these costs for a variety of pipe sizes must be computed to enable the optimum size to be selected for a given case.

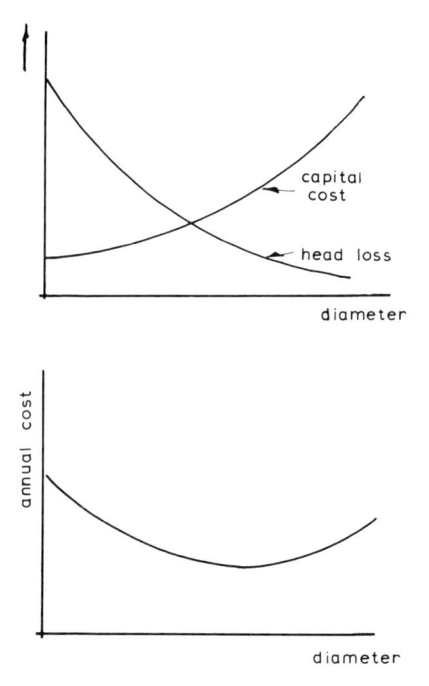

FIGURE 42. Optimum size of a pumping main

The layout of water mains is greatly dependent on local conditions and topography but, in general, the system is arranged in a series of pressure zones, each with a range of 40 to 60 m head, to prevent excessive pressures at any point in the system and the consequential loss of water by leakage. Dead-ends in the distribution network are undesirable since they are likely to cause taste and odour problems. Ring mains obviate the need for dead-ends and are able to reduce the area of loss of supply in the event of a burst. Cross-connections between zones which can be opened in an emergency are also useful in this respect.

VALVES ON MAINS

Water mains are normally laid to follow ground contours with a minimum cover of about 1 m so that provision must be made for draining each section of the pipe and for the entry and escape of air (Fig. 43). A washout valve is provided at each low point in a main to permit removal of debris after cleaning of the line which may need to be an annual operation with some waters. A removable length of pipe is usually provided for the insertion of cleaning devices and for the removal of these devices and the debris which they force along the pipe. At high points air collects as the line is filled and an airlock may result. During normal operation, air may come out of solution due to pressure variations and again this air collects at high points. If a pipeline is suddenly emptied by a burst, partial vacuum conditions could occur causing the pipe to collapse unless the vacuum were released. An air valve (Fig. 44) at each high point on the line prevents these troubles. The small orifice releases air at normal working pressure since air entering the chamber forces the rubber ball away from its seat. Normal working pressure keeps the vulcanite ball tight against the large orifice but it is capable of releasing large volumes of air during filling of the pipe and also of admitting large amounts if the pipe is suddenly emptied.

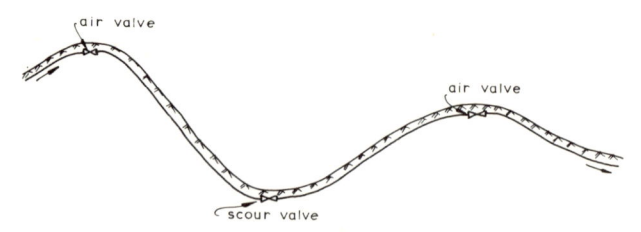

FIGURE 43. Positioning of air and scour valves on a pipeline

Automatic self-closing valves are often fitted in trunk mains to shut off the flow if a specified discharge is exceeded on the assumption that the large flow must be due to a burst pipe. Isolating and flow control valves are usually of the sluice valve type with a sliding metal plate operated by a screwed spindle. When fully open they provide a full-bore passage through the valve which permits the use of cleaning 'pigs' in the pipeline. These cleaning

Distribution of Water

devices, which may be in the form of scrapers, brushes or foamed plastics swabs, are slid into the pipeline using a removable length of pipe. They are forced along the pipe by the water pressure behind them.

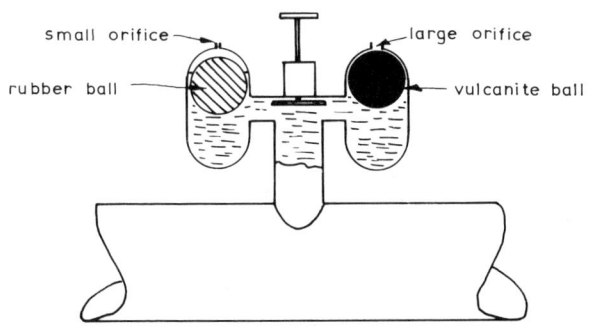

FIGURE 44. Diagrammatic arrangement of an air valve

PUMPS

Electrically-driven centrifugal pumps are normally used for water supply purposes (Plate 21) and may be installed in a dry well alongside a wet suction well with the motors and control gear above ground level as a protection against flooding (Fig. 45). Alternatively, submersible pumps may be used, particularly for borehole supplies. In this case everything except the electrical control gear is below water. With deep boreholes multi-stage pumps are necessary to lift the water to the surface (Plate 22).

Because of the need to maintain the supply of water, alternative electricity feeders from different power stations or automatically-started diesel generators are normally provided on major installations so that breakdown in the public electricity supply does not shut down pumping stations and treatment plants. Most mechanical plant in a waterworks is duplicated so that failure of one unit does not imperil the production of potable water.

SURGE PREVENTION

Severe surges can be caused in pipelines by the sudden operation of a valve or sudden starting or stopping of a pump. Considerable pressures can be generated or, conversely, high

PLATE 21. A vertical spindle centrifugal pump with the electric motor installed above the pump

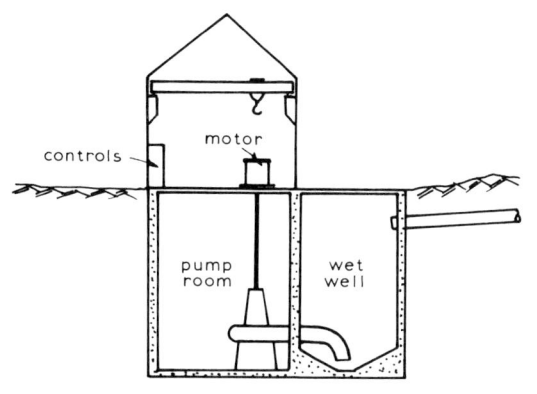

FIGURE 45. Typical pumping station

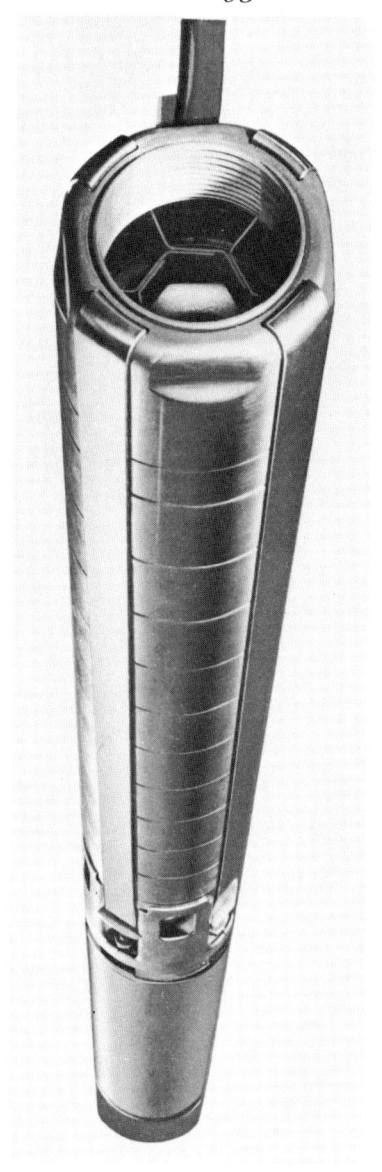

PLATE 22. An electrically driven submersible pump for borehole supplies

vacuums can be produced. In either case major damage to the pipeline may occur.

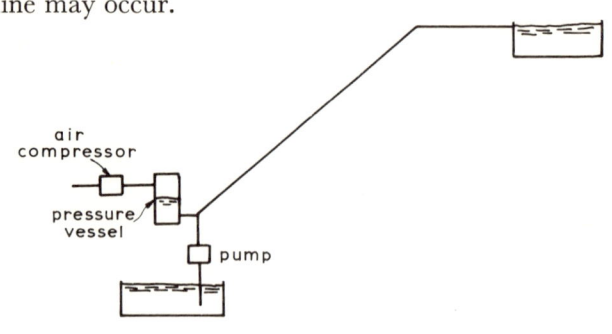

FIGURE 46. Air-loaded vessel for surge prevention

Surge prevention can be achieved by a variety of methods of which the air vessel or surface-feeder tanks are the most satisfactory. An air-loaded vessel (Fig. 46) connected to the pump delivery will maintain a diminishing flow along the main in the event of a sudden pump failure, thus limiting the resulting pressure and velocity surges in the system. The air vessel receives water as the pump starts up and thus damps out pipeline surges. Surface feeder tanks (Fig. 47) spaced approximately 10 m vertically on the hydraulic gradient and connected to the main by non-return valves will also control surges but provision must be made for filling and emptying the tanks. Other methods of surge control include slow-closing motorized valves, relief by-pass valves and flywheels on pumps. Whichever method is used a complex mathematical analysis is necessary to enable a satisfactory system to be designed.

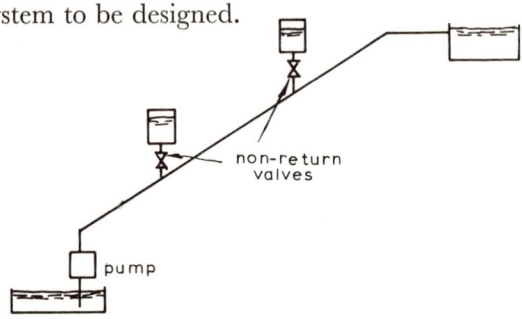

FIGURE 47. Surface feeder tanks for surge prevention

CORROSION

Because of the high capital cost of water mains and the inconvenience of having corrosion products entering the water, effective corrosion control measures are essential. Metal pipes are hot dipped, or lined internally by spinning, with a bituminous compound which gives good protection provided that it completely covers the surface of the pipe. Bituminous compounds are rather soft and the coatings are easily damaged during laying and jointing operations. Welding destroys the coating and it must be made good both internally and externally. Concrete linings may be spun into steel or iron pipes to give a strong although somewhat porous lining which again must be made good after jointing has been completed.

Internal protection of metal pipes can also be provided by chemical conditioning of the water which flows through them. Polyphosphates combine with ferric oxide to form a glassy phosphate film which prevents further corrosion. A residual concentration of about 0·5 mg/l phosphate is necessary to maintain the film. Alternatively, a hard protective coating of calcium carbonate can be formed by the controlled addition of calcium to the water, the dose being calculated from the Langelier saturation index. Chemical conditioning has the disadvantage that the coating may not be uniform and some sections may be unprotected. The chemicals used may be undesirable for certain industrial uses of water although they are quite acceptable in potable supplies.

External protection of pipelines is, of course, equally important since many soils are aggressive. The outside of metal pipes is usually protected by a sheath of hessian dipped in bituminous compound and wrapped tightly around the pipe, again taking care to ensure continuity of protection at joints. Cathodic protection is an effective way of preventing external corrosion of metal pipes either with sacrificial anodes or with an impressed current (Fig. 48). The aim is to produce an electrical cell with the pipe as the cathode which does not corrode. For sacrificial anodes, metals such as magnesium or aluminium—which are higher in the electrochemical table than iron—are used. The anodes corrode and must be replaced as necessary to maintain protection of the iron. With an impressed current the anode may be scrap-iron, which is cheaper than the other metals, but an external source of power is required.

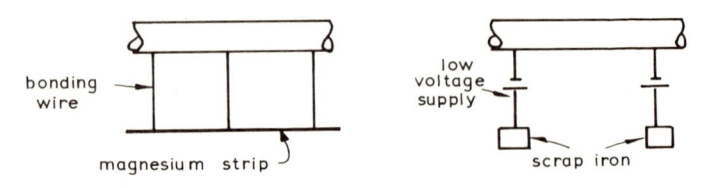

FIGURE 48. Cathodic protection

SERVICE RESERVOIRS AND WATER TOWERS

It is usually desirable to operate treatment plants and pumping stations at a steady output since this simplifies control. The demand for water varies throughout the day so that it is almost universal to provide storage of treated water in service reservoirs, or water towers if suitably elevated natural sites are not available. It is then possible to have a constant inflow to the service reservoir and use its storage capacity to absorb excess flow during the night and to provide additional flow from storage during the day when demand exceeds the inflow. (Fig. 49). Service reservoirs also provide a reserve of water in cases of treatment plant or pumping station breakdown and the storage may be useful for fire-fighting purposes. Very often they are used as break-pressure

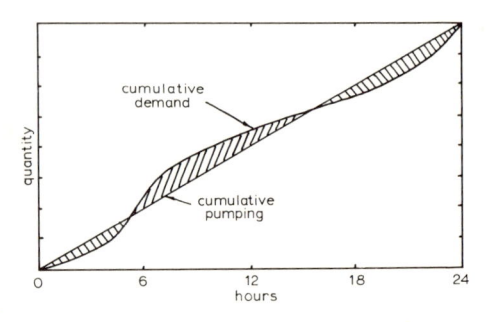

FIGURE 49. Function of a service reservoir. Storage, indicated by the hatched areas, enables a fluctuating demand to be provided from a constant rate supply

points to destroy excess pressure when dividing the area of supply into pressure zones.

Service reservoirs and water towers store treated water and are always covered to prevent aerial pollution from birds and windblown material. Their construction is such that contaminated surface- or groundwater cannot gain access to the interior. Modern service reservoirs are usually partitioned to permit cleaning and maintenance to be carried out without taking the whole reservoir out of supply. Water towers are usually of rather smaller capacity than reservoirs because of the high cost of elevated water-retaining structures. Whilst some towers tend to be rather utilitarian in appearance (Plate 23), several aesthetically pleasing towers have been built in recent years (Plate 24).

PREVENTION OF LEAKAGE AND WASTE

Excessive amounts of water may be lost from a water supply system by leakage from reservoirs, mains and domestic fittings as well as by waste on the part of consumers. As a general rule, in the UK, a leakage and waste component of about 10 per cent of the total demand is accepted by many undertakings. This may seem a large amount but it is usually made up of a multitude of small leaks and at the present value of water it is uneconomic to find and rectify all such leaks.

Leak and waste detection in a water distribution system is not an easy matter because of the complexity of such networks. As an initial step, rate of flow meters may be temporarily installed in mains and their records inspected to show excessive night flows after deduction of known industrial demands. The area served by the particular main is then examined in more detail with visits to premises to fit new washers to leaking taps, etc. All valves in the section are checked and stethoscopes are used to detect the sound of water escaping under pressure. Other methods employed to find leaks include the use of radioisotopes and halogen gases which leave detectable residues in the vicinity of significant leaks.

PLATE 23. A conventional reinforced concrete water tower

PLATE 24. A modern reinforced concrete water tower of pleasing appearance

7 Collection of Wastewater

The majority of water supplied to consumers is discharged from the premises as domestic sewage or industrial effluent and it is necessary to provide a sewerage system to convey these discharges to a treatment plant or disposal point. Urbanization of an area radically alters its run-off characteristics, since paved areas, roads and roofs are relatively impermeable compared with the grassland of an undeveloped site. Flooding of farmland and pasture may be tolerated at quite frequent intervals but when the land is developed, flooding would occur more frequently due to the greater run-off caused by the higher impermeability. The nuisance and damage caused by the flooding would, in any case, become unacceptable. The surfacewater run-off from an urban area must also, therefore, be conveyed in sewers to a suitable point of discharge. The volume of surfacewater from even a moderate rainfall is greatly in excess of the dry weather flow (dwf) of foul sewage from a built-up area and this poses certain design problems.

Older communities are usually sewered on the combined system in which foul sewage and all the surfacewater run-off from the area is collected and transported in a single system of sewers. Such sewers are difficult to design because of the wide range of flows experienced, e.g., 0·4 dwf at night to perhaps 100 dwf after a heavy storm, and it is usually impractical to convey all of the flow to the treatment plant. It has thus become common practice to install stormwater overflows (Fig. 50) at intervals along the sewer to limit the maximum flow to between 6 and 12 dwf, any excess flow being discharged to the nearest watercourse where it may often cause considerable pollution. Normally, of course, the receiving watercourse will contain a high flow because of the rainfall so that the discharge of storm sewage is somewhat diluted. Stormwater overflows are, however, clearly undesirable particularly in urban surroundings where they can cause aesthetic offence as well as producing gross pollution from the first flush of storm sewage discharge.

Because of the undesirable features of stormwater overflows most new developments are sewered on the separate system with different sewers for foul sewage and surfacewater run-off. The foul sewers have no overflows and run directly to the treatment plant, whereas the storm sewers which carry only surfacewater can discharge to any suitable watercourse without producing a pollution problem. Care must be taken to ensure that the watercourse is capable of carrying the stormwater discharge without causing flooding downstream. Hydraulic design of the sewers is made easier by the adoption of the separate system and, although two pipes must be installed, the extra cost over a combined system need not be so much greater since much of the cost is for excavation (it is usual to place both pipes in the same trench for much of their length). The adoption of separate sewerage systems plays an important part in the reduction of river pollution in urban areas. Unfortunately, since many of our older communities have combined sewerage systems it would be difficult and costly to convert these areas to the separate system. Never-

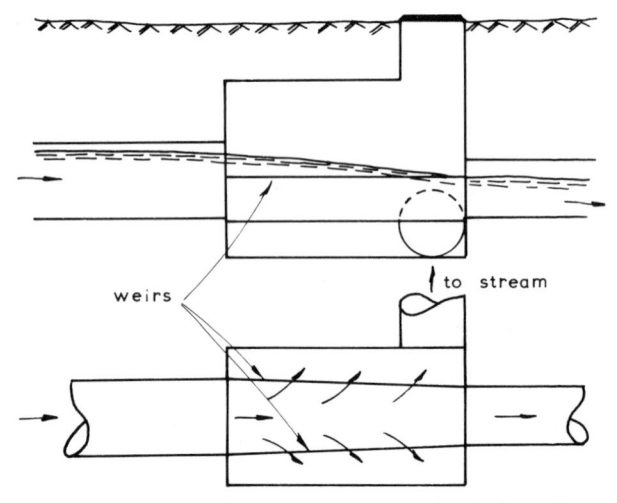

FIGURE 50. Storm water overflow. Flows in excess of 9 dwf are discharged over side weirs at a level such that the flow in the downstream section of the sewer is restricted to 9 dwf

Collection of Wastewater

theless, when redevelopment work is carried out in a city the opportunity is often taken to install separate sewers as part of a long-term conversion of the drainage system.

SEWER DESIGN

The main characteristics of sewers are, that, in most cases, they operate under gravity flow conditions (some exceptions being inverted syphons under rivers or pumping mains from low-lying areas), the flow in them is unsteady and non-uniform, and the sewage they transport contains relatively large amounts of floating and suspended solids. This last point is particularly important since it is undesirable for suspended matter to be deposited in the sewer. Road grit, which is a common constituent of the flow in combined sewers, may cause blockages and organic solids tend to be stranded on the edge of the water surface as the flow falls. Deposits of organic matter decompose anaerobically with the production of hydrogen sulphide. Hydrogen sulphide dissolves in any moisture present on the walls and roof (soffit) of the sewer and it may then be utilized by bacteria to produce sulphuric acid which can cause corrosion.

Where possible, sewers are designed to be self-cleansing at flows of 2 dwf so that cleansing should occur at least once per day. It is also necessary to limit the maximum velocity so that the sewer fabric is not damaged by abrasion, bearing in mind that a

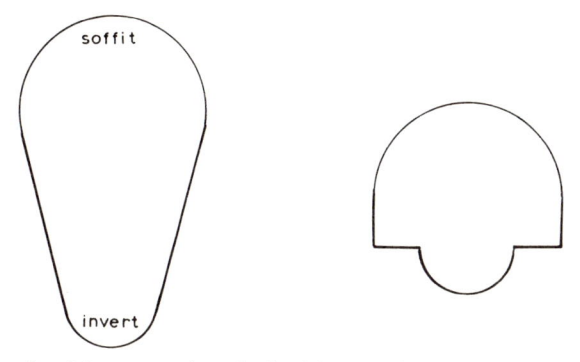

FIGURE 51. Special sewer sections. Both of these sections are attempts to provide a small pipe to give suitable velocities in dry weather whilst also giving adequate capacity for conveying storm water discharges

suspension of sand and grit is an efficient scouring fluid. In order to satisfy these criteria it is usual to keep velocities in the range 0·5 to 3·5 m/s at dwf and maximum flow respectively. With combined sewers it may not be possible to comply with both velocity criteria and special pipe sections are sometimes used to give the desired hydraulic characteristics (Fig. 51). The low flows are contained in the bottom (invert) of the sewer, the larger cross-section coming into use as the flow increases.

For much of the time, sewers flow only partly full although in times of heavy rain, excess run-off may cause combined sewers to surcharge and run as pressure pipes with sewage backed up in the manholes. In extreme cases, the manhole covers may be blown off by the excess pressure and once the head of sewage reaches ground level flooding will occur. Because of the variable nature of the flow in sewers and the limited information about their roughness after some years of service, flow calculations are often based on empirical formulae such as that developed by Crimp and Bruges from the work by Manning:

$$v = kr^{2/3}s^{1/2}$$

where k = a constant depending on pipe roughness;

r = hydraulic radius $\left(\dfrac{\text{cross-sectional area of flow}}{\text{wetted perimeter}}\right)$;

s = slope of invert.

A more comprehensive method of design uses charts (devised at the Hydraulics Research Station), based on the Colebrook and White formula.

Small-diameter sewers are usually made of salt-glazed fireclay or pitch-fibre pipes, with flexible joints made with rubber rings so that ground movements do not open the joints and allow leakage. Such flexible joints are much easier to make than the traditional mortar joints. For sewers larger than about 300 mm diameter, concrete pipes are usually adopted employing ogee, spigot and socket or flexible joints as depicted in Fig. 52. With large sewers or where sewers are laid under roads, or at great depths, concrete surrounds are used to provide extra strength since considerable loads arise from the backfilling in deep

Collection of Wastewater

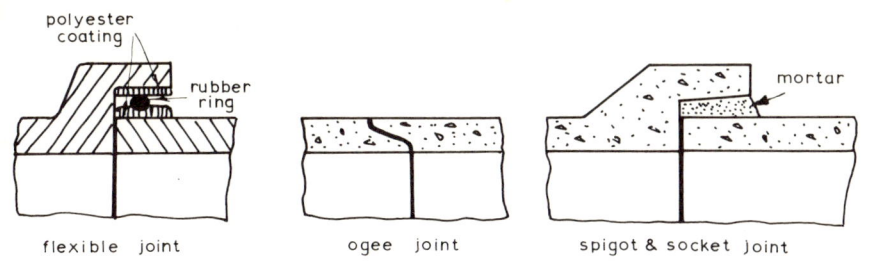

polyester coating / rubber ring

flexible joint · ogee joint · spigot & socket joint (mortar)

FIGURE 52. Joints in sewer pipes

PLATE 25. A large sewer under construction. Precast concrete segments are inserted as excavation proceeds to make complete rings. A lining of glazed engineering bricks completes the sewer and provides a smooth and durable surface

trenches. Large sewers are usually constructed from cast *in situ* or precast concrete segments lined with engineering brick to provide higher resistance to abrasion and corrosion (Plate 25). In special circumstances, where the flow is under pressure in syphons or pumping mains or where there are very high velocities, cast-iron pipes may be necessary.

Many old sewage pumping stations contained magnificent steam engines but modern stations normally use electrically-driven centrifugal pumps. Specially designed impellers enable them to pass large solids without seriously reducing pump efficiency. Complete factory-built sewage pumping stations are now available and only require to be placed in an excavated hole and connected up to sewers and power supply to become fully operational.

A sewerage system is usually based on the street layout with a lateral sewer down each street, the flow in each lateral being collected by an intercepting sewer (Fig. 53). Manholes are placed at each change of direction to permit inspection and cleaning to take place and to give access to larger sewers. The maximum distance between manholes is about 120 m so that the distance which a worker in the sewer has to walk to an exit is always relatively short. This is an important point since sudden rainfall can produce a very rapid rise in flow, sufficient to sweep a man off his feet, if not engulf him. Because of the possible presence of hydrogen sulphide in sewers from anaerobic decomposition of deposited organic matter, ventilation is essential. This is normally achieved via house connection soil pipes supplemented by ventilated manhole covers. Care is always necessary when entering a sewer because toxic concentrations of hydrogen sulphide may be present and a number of fatalities have resulted from such situations. Rats infected with Weil's disease are often found in large sewers and the disease can be transmitted to man by

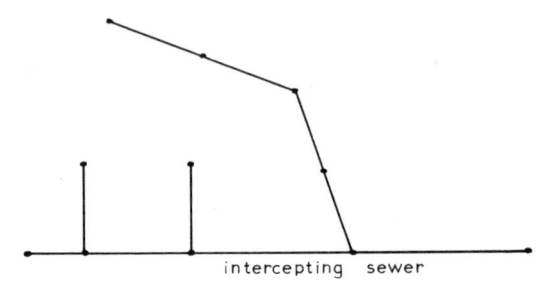

FIGURE 53. Basic elements of a sewerage system. Laterals are connected to an intercepting sewer, manholes are placed at each change of direction and at a spacing not exceeding 120 m

Collection of Wastewater

contact with sewage containing their urine. Because of these hazards, work is closely supervised and carried out by specially-trained men provided with breathing apparatus when necessary.

SURFACEWATER DRAINAGE

When rain falls some will infiltrate into the ground—the amount depending on its nature and moisture content—and some will appear as surface run-off. In undeveloped land, streams and ditches provide natural surfacewater drains so that flooding may only occur during heavy storms. When relatively

Table 12. Typical impermeability factors

Surface	Impermeability factor
Watertight roof	0·70 — 0·95
Asphalt	0·85 — 0·90
Concrete flags	0·50 — 0·85
Macadam	0·25 — 0·60
Gravel	0·15 — 0·20
Grass (level)	0·10 — 0·20

permeable grassland is replaced by a housing development with more-or-less impermeable roofs, roads and paved areas, the run-off is much increased and the inconvenience and expense of flooding becomes unacceptable.

The proportion of the precipitation which appears as surface run-off is governed by the impermeability of the area and some typical impermeability factors are given in Table 12. During a storm the impermeability factor is not constant since at the beginning raindrops lie on the surface without running off and do not move until sufficient rain has accumulated to overcome surface tension. Run-off will also tend to continue for some time after rainfall has ceased because of the amount of water stored on the surface. The impermeability will also depend on the previous rainfall history, since if the surface is still wet from an earlier storm a higher impermeability would be recorded during the second storm. An extreme example of this point would be grassland which normally has a low impermeability but, when saturated after heavy rain, may be almost impermeable and would thus result in flooding to a larger extent than would have been estimated from the particular rainfall intensity.

Run-off from a drainage area is given by:

$$Q = 0.278 A_p R$$

where Q = run-off m³/s;
 A_p = area (km²) × impermeability factor;
 R = rainfall intensity mm/h.

As described in Chapter 4, rainfall intensity depends on the duration of the storm and the return period adopted. For design purposes it is necessary to select the duration of the storm on the basis of catchment characteristics and the return period by taking into account the type of development in the catchment, including consideration of the economic effects and possible hazard to life due to flooding.

Figure 54 shows a sewered catchment over the whole of which rain starts to fall at the same time. The flow at the exit from the area will begin almost at once due to run-off from rain on areas in the immediate vicinity. As time passes, the flow increases as more and more run-off reaches the exit. Eventually, the whole area is contributing run-off at a maximum discharge equal to the impermeable area run-off. When the storm ceases the flow will decrease as run-off from the nearest parts of the catchment stops almost at once but flow from the more distant parts will continue for some time. The time taken to reach the maximum discharge is known as the time of concentration for the catchment and is

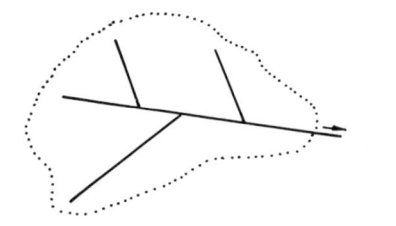

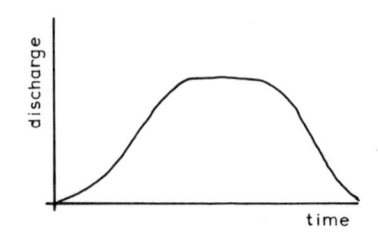

FIGURE 54. A surfacewater sewerage system and its discharge hydrograph

Collection of Wastewater

made up of the time of entry (the time taken for rainwater to flow via gutters and gulleys to the actual sewer, usually 2 to 5 min in built-up areas) and the time of flow in the sewer (normally calculated on the basis that all sewers are running full). It can be shown that the maximum run-off from an area occurs when the storm has a duration equal to the time of concentration of that area.

The intensity of rainfall is obtained from the Bilham formula (page 32) using the appropriate return period, commonly one year but possibly longer, where flooding would be particularly costly or undesirable. Alternatively, the Ministry of Health rainfall formula, which approximates to a one year return period, may be used:

$$R = \frac{760}{t+10} \quad \text{for } t \leqslant 10 \qquad R = \frac{1020}{t+20} \quad \text{for } t > 10$$

where R = rainfall intensity mm/h;
       t = duration of storm min.

A time–area diagram (Fig. 55), which shows the impermeable area contributing at any moment in time after the start of a storm, is often useful in illustrating the run-off characteristics of a drainage area.

A newer and more rational method of design suitable for large drainage areas has been developed by the Road Research Laboratory. It involves the construction of a run-off hydrograph

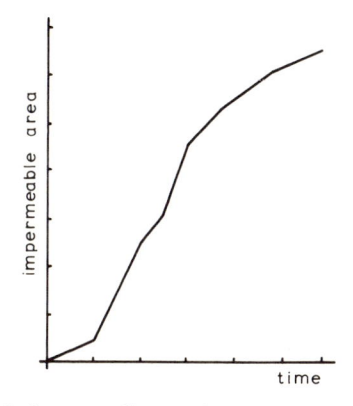

FIGURE 55. A time-area diagram for a surfacewater sewerage system

for the catchment which, if derived directly from rainfall measurements and the impermeable areas, does not reproduce the measured run-off observed in actual drainage areas. This discrepancy is due to the storage capacity present in the sewers which is not allowed for in a conventional time-area diagram in the plotting of which it is assumed that the velocity of flow in the sewers is constant throughout a storm. Since this would mean that the same depth of flow exists in each sewer throughout a storm the assumption is clearly incorrect. Initially, the depth of water and hence the velocity will be small with a gradual increase taking place until the maximum discharge is reached. By the time this condition has been attained a considerable volume of water will have been stored in the system. It is, thus, necessary to modify the calculated run-off hydrograph to allow for storage. For any given sewer system there is a relationship between the depth of flow at the outlet and the total volume of water retained in the system, which can be obtained by examination of existing discharge records or by computation. Knowing the storage/discharge relationship for the system it is possible to produce a corrected hydrograph by subtracting the quantity in storage from the calculated hydrograph based on rainfall intensity and impermeable areas. Because of the complexity of the calculations a computer solution is necessary and the savings in cost produced from the use of this design method only become significant with systems containing sewers larger than about 600 mm diameter.

8 Water Pollution

As explained in Chapter 2, all natural waters are impure to some degree in that they contain constituents other than H_2O. The term pollution is, however, usually reserved for the situation where the presence of contaminants in the water leads to the development of undesirable characteristics. It should not be thought that all such pollution is man-made since, although much pollution is the result of wastewater discharges, natural pollution by run-off, decaying vegetation etc., is sometimes quite important.

SURFACEWATER POLLUTION

A body of natural water supports a complex ecological system (Fig. 3), the type and magnitude of which depends on the composition and properties of the water. Thus, in a mountain stream conveying run-off from a rocky ground surface the amount of organic matter in solution will be small so that the water will only be able to support a relatively small amount of life. Unless organic food and the lower forms of life exist in significant concentrations the higher forms of life, such as fish, cannot proliferate. Lowland rivers are likely to have more prolific growths than upland waters because of the higher food concentrations in the water due to natural run-off from agricultural land as well as from effluent discharges, the actual source being relatively unimportant. Indeed, there have been cases where a limited amount of sewage pollution in a river has markedly improved the fishing due to the greater availability of food. In a stream with little organic matter in solution there are only small numbers of micro-organisms present although many different species will be represented. As the concentration of organic matter increases so does the number of micro-organisms in order to maintain a balanced ecosystem. Under normal conditions, the organic matter is broken down aerobically since, although the dissolved oxygen present in the water is consumed during the oxidation reactions, it is replenished by oxygen transferred from the atmosphere. In the presence of excessive amounts of organic matter, or if oxygen transfer is hindered, the oxidation processes will consume oxygen more rapidly than it can be replenished. In these circumstances the dissolved oxygen in the water becomes depleted, resulting in the death of the higher forms of life which are sensitive to oxygen levels and thus unbalancing the biological population in the water. Large numbers of bacteria then produce anaerobic conditions and obvious signs of gross pollution occur. When a water becomes anaerobic, only the simple microorganisms such as bacteria can live and it becomes turbid and unattractive with added risk of considerable amounts of hydrogen sulphide and methane, the end products of anaerobic reactions, being manufactured. Given sufficient time, remembering that anaerobic reactions are slow, stabilization of the organic matter will proceed and self-purification will eventually result in a return to a stable water with ample dissolved oxygen and a balanced community.

The oxygen balance of a water is of great importance in controlling the effects of pollution and, because of the variable oxygenation characteristics of surfacewater each case must be considered separately. It is impossible to say that a given BOD load discharged to a certain volume of water will result in a particular degree of deoxygenation. When considering effluent discharges, constituents which affect the oxygen balance of the receiving water are usually of the greatest significance. Apart from obvious oxygen-consuming materials, such as organic compounds and inorganic reducing agents which are oxidized in the water, it should be appreciated that substances like oils and detergents can hinder the transfer of oxygen from the atmosphere by forming protective surface films. In addition, heated discharges, e.g., cooling water, can significantly affect the oxygen balance because the saturation level of dissolved oxygen is reduced as temperature is increased, e.g., 14·6 mg/l at 0°C, 9·1 mg/l at 20°C, 7·6 mg/l at 30°C (in fresh water, the concentrations being somewhat lower in salt water). As the temperature is increased, fish require more oxygen for their respiration but, since the increase in temperature reduces the available oxygen in the water, it will be appreciated that discharges of heated water can be directly responsible for fish being killed.

Pollution does, of course, have effects other than on the oxygen balance of the receiving water. Toxic substances, e.g., heavy metals from metal finishing wastes may reduce the amount of aquatic life or even produce virtually barren streams. Certain algae produce toxins which can make water unsuitable for consumption by cattle (and, presumably, by man). The increasing range of exotic chemicals produced by modern industry and now being found in effluent discharges requires a careful watch for possible toxic effects on natural stream life and also on man if the water is abstracted for public consumption. Even inert solids such as china clay particles can be indirectly toxic to fish if present in high concentrations since they blanket the stream bed cutting off the supply of food for the fish which eventually die of hunger or migrate to a healthier area.

Apart from the possible danger to public health, pollution of water is clearly undesirable for many reasons, some of which are given as follows:

(a) Contamination of water supplies resulting in an additional load on water-treatment plants, increasing the cost of treatment and often causing taste and odour problems. Sewage and many industrial wastes contain inorganic salts so that their discharge results in a build-up of dissolved solids in the receiving water which may eventually reach an excessive concentration. Conventional methods of water treatment do not remove significant amounts of dissolved solids so that such a water could become untreatable in a normal water purification plant.

(b) Restriction of recreational use. Bathing and swimming are unwise in polluted waters and, indeed, in some parts of the world, notably certain American states, bathing waters have to satisfy bacteriological quality standards. There is, however, no rational basis for such standards. When gross pollution occurs, even boating becomes unpleasant and the actual amenity value of the water is much reduced.

(c) Effect on fish. Game fish (trout and salmon) require high DO levels and are not normally found if the DO is less than 5 mg/l. Coarse fish can exist at DO levels down to about 2 mg/l but at such levels they are likely to be sensitive to the presence of other contaminants in the water. Because of their sensitivity to pollution, fish very often provide the first indication of pollution and there are some lengths of rivers in the UK which are so polluted that they cannot support any fish life and much longer lengths can only provide coarse fishing. The Tame, which drains the Birmingham conurbation, is an unfortunate example of a fishless river and such rivers as the Tyne, now grossly polluted were full of salmon before the industrial revolution. On a brighter note, it should be recorded that extensive pollution prevention measures on the Thames have resulted in the return of many species of fish which had previously been absent from the lower reaches. Angling is, of course, a major hobby with a large industry manufacturing equipment. Commercial salmon fishing in the British Isles has an annual turnover of about £1 million sterling.

(d) Creation of nuisances by appearance and odour following gross pollution (Plate 26). Prolific growths of sewage fungus become most unsightly and if a water is anaerobic its black appearance and the bubbles of gas which issue from the surface are most unattractive. Debris stranded on the bank is visually objectionable and may constitute a health hazard to children playing there. The odours from anaerobic water, mainly hydrogen sulphide with its characteristic rotten egg smell, are sometimes overpowering and the gas can damage paint work on nearby structures. The foam produced by detergents is unsightly and most noticeable at weirs and other points of turbulence. When synthetic detergents were first introduced it was found that they were not easily broken down by sewage treatment processes with the result that much of the detergent entering the sewage works was discharged in the effluent. Large banks of foam appeared in rivers and the detergent content hindered oxygen transfer. Detergent manufacturers were persuaded to adopt new 'soft' formulations which were more easily broken down biologically so that the problem was greatly reduced. Pressure is now being placed on industrial organizations to switch to the use of soft detergents.

(e) The deposition of suspended solids may form sludge banks which can become a hindrance to navigation and necessitate regular dredging to maintain shipping channels.

PLATE 26. Prolific growth of plant life 'sewage fungus' in a stream subject to organic pollution. The bed of the stream and many of the stones are covered with plant growth

THE OXYGEN BALANCE

When organic matter enters water containing dissolved oxygen it will be broken down aerobically by micro-organisms and an indication of the amount of oxygen required is given by the BOD test in which it is assumed that the rate of oxidation is proportional to the amount of organic matter remaining, i.e., a concentration dependent reaction which can be expressed as:

$$BOD_t = L(1 - 10^{-k_1 t})$$

where BOD_t = BOD after time t;
L = ultimate BOD;
k_1 = reaction rate constant which depends on the nature of the organic matter, temperature and other environmental factors.

For the standard conditions of test, five days at 20°C, domestic sewage has a k_1 value of about 0·17 d^{-1} and the five-day BOD is about two thirds of the ultimate BOD, which would be reached theoretically after about 20 days. The oxidation of nitrogen compounds to nitrate (nitrification) also consumes oxygen so that measured BOD values may be higher than would be recorded from oxidation of the carbonaceous matter alone. It is, however, possible to prevent nitrification in the BOD test by the addition of thiourea which inhibits the growth of nitrifying bacteria without interfering with the activities of the bacteria responsible for the oxidation of organic carbon compounds.

A natural surfacewater is usually saturated with oxygen but, as soon as BOD is exerted as the result of an effluent discharge, the DO falls and an oxygen deficit is created in the water. The existence of this oxygen deficiency causes oxygen to be transferred from the atmosphere to the water in an attempt to bring the DO back to the saturation level. The rate of solution of oxygen is proportional to the oxygen deficit, i.e.,

$$D_t = D_a 10^{-k_2 t}$$

where D_t = oxygen deficit after time t;
D_a = initial oxygen deficit;
k_2 = re-aeration constant.

The re-aeration constant depends on the nature of the flow and the channel configuration and is also temperature dependent. A mountain stream may have a re-aeration constant a thousand times greater than that for a stagnant lake and would thus be able to assimilate a much larger BOD load per unit volume without producing serious oxygen depletion.

As the amount of oxygen consumed by BOD uptake rises the

oxygen deficit increases so that the initially slow rate of re-aeration is accelerated and, if the pollution was not excessive, a critical point is eventually reached at which the rates of deoxygenation and re-aeration become equal. Thereafter, re-aeration is able to supply more oxygen than is being consumed by the BOD reaction and thus the DO will eventually return to saturation. This situation results in the classical oxygen sag curve (Fig. 56) which is derived from consideration of the differences in rate of oxygen uptake and re-aeration.

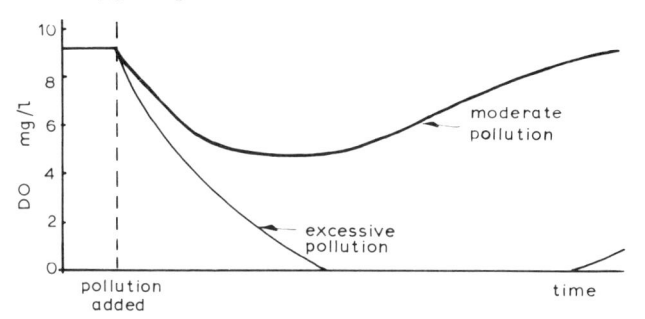

FIGURE 56. The dissolved oxygen sag curve. With moderate pollution the dissolved oxygen levels remain high enough to support fish life. Following excessive pollution all dissolved oxygen is removed from the water and the river becomes almost barren and obviously polluted

The sag curve may be expressed mathematically as

$$D_t = \frac{k_1 L_a}{k_2 - k_1} \left(10^{-k_1 t} - 10^{-k_2 t}\right) + D_a 10^{-k_2 t}$$

where L_a = ultimate BOD initially, and other symbols are as before.

The sag formula, by Streeter and Phelps, is based on the assumptions that the only factors affecting the oxygen balance are uptake of oxygen by BOD and re-aeration from the atmosphere, and that there is no alteration in the flow or added pollution in the stretch of river under consideration. In reality, many other factors such as surface run-off, plant photosynthesis and the effect of mud deposits can significantly affect the oxygen balance of a water. The sag concept was developed from studies on large American rivers and it has only limited application to the much shorter rivers found in the UK.

TOXIC MATERIALS

It is often assumed that if fish can live in a particular water it cannot contain matter which would be toxic to man. This assumption has never been proved. In any case, it should be realized that fish will certainly be found in waters containing micro-organisms which would be harmful to man. It does not, therefore, follow that a water supporting fish is safe to drink without treatment. Fish can be kept in the treated effluent from a well-operated sewage works but the drinking of such an effluent without further treatment by man would be most unwise. In considering the effects of toxic matter it is necessary to distinguish between the short-term effects of relatively large concentrations and the possible long-term effects arising from the continued ingestion of very small concentrations of toxic matter over many years. Most poisons have a certain threshold concentration below which the material is harmless but this does not apply to cumulative poisons, like lead, or to radioactivity which produces genetic damage even at very low levels of exposure.

The establishment of concentrations of contaminants which are toxic to fish is difficult because of the effects of environmental factors such as pH, temperature, other constituents of the water, the species of fish and the duration of exposure. Most published data on toxic levels can thus only be taken as a guide and they are normally expressed as median tolerance limits (TL_m) or 50 per cent lethal doses (LD_{50}). These are the concentrations of material under investigation at which 50 per cent of the fish exposed are able to survive for the period of test, which is usually 48 or 96 hours. Typical TL_m values are 0·02 mg/l for aldrin (an insecticide), 0·05 mg/l for cyanide and 2 mg/l for ammonia. Permissible levels in rivers are commonly set at 10 per cent of the TL_m values. The Water Pollution Research Laboratory maintains an information service on toxic chemicals (INSTAB) which is able to provide details of the toxicity of numerous substances under a variety of conditions. In the future, much more use will be made of lowland rivers as a source of domestic water and the possible toxicity of discharges from industrial premises upstream is of considerable concern. The identification of contaminants, particularly at low concentrations, is often difficult and it is not possible at the present time to devise a monitoring system which would give rapid warning of the presence of all toxic materials which might be present in the raw water. It is thus of the utmost importance that the discharge of potentially toxic wastes to rivers used for water supply purposes should be carefully controlled. The accidental discharge of toxic chemicals (oil, etc.) is an ever-present hazard in river systems. An accident involving a tanker carrying chemicals may result in the discharge of large volumes of dangerous liquid which often finds its way into watercourses. The hazards of such situations can only be reduced by the rapid reporting of the accident to the appropriate river authority and water undertakings. Even then it may be impossible to prevent damage to the fish and bird life in the river.

EUTROPHICATION OF LAKES

All natural lakes undergo a slow change from low nutrient content waters (oligotrophic) to nutrient rich waters (eutrophic). The process of eutrophication is, in fact, an enrichment of all forms of nutrients in the water but it is often taken to mean the build-up of nitrogen and phosphorus which directly control the growth of algae and other aquatic plants. Under natural conditions, eutrophication is a very gradual process, taking perhaps hundreds of years to change the character of a lake significantly in isolated and barren surroundings. In an increasingly large number of areas, however, the process is being accelerated by human activities, such as farming, with intensive use of nitrogen and phosphorus fertilizers and by effluent discharges which contain these elements. As a result, some lakes, notably certain of the Great Lakes in North America and some in Italy and Switzerland, have radically altered in character in a short period of time. Excessive growths of weeds and algae have resulted in water treatment problems and interfered with recreational use. Fish may be killed by oxygen depletion at night by the large numbers of algae and also because dead algal cells consume oxygen as they sink to the bottom. High nitrate concentrations have made the treated water unsuitable for consumption by babies because of the risk of methaemoglobinaemia. In Lakes Erie and Ontario, which receive the effluents from large populations, the chloride content has increased by 500 per cent in the last 50 years. In Lake Erie the ammonia content has increased by 600 per cent in 40 years and, as a result of the greater biological activity in the lake, large DO deficits occur much more often and over larger areas of the lake than previously. In the UK, lakes such as Windermere are showing signs of accelerated eutrophication due to effluent discharges from the ever-increasing population along their shores. Windermere has recently been brought into use as an additional source of water for Manchester and, because of the discharges into it, much more rigorous treatment is required than has been satisfactory for the more isolated reservoirs which originally supplied all of Manchester's needs.

The realization that limitation of BOD and SS in effluents was not always sufficient to prevent undesirable consequences following their discharge has resulted in the consideration of partial or complete removal of nutrients. Such nutrient stripping usually requires an additional stage of treatment at extra cost. Much of the available evidence on algal growths in lakes indicates that phosphorus is the critical element in determining biological activity in a body of water. Most of the lakes which are known to produce troublesome blooms of algae have phosphorus levels greater than 0·1 mg/l as organic phosphorus. However, there is evidence that the critical level with inorganic phosphorus is much lower, perhaps 0·005 mg/l. Nitrogen is also important for algal growth and it has been suggested that the critical level is around 0·3 mg/l. Nitrogen can, however, be obtained by fixation from the atmosphere so that it is generally the phosphorus levels which control growths.

Synthetic detergents contain phosphates and contribute about half of the phosphorus found in sewage (most of the remainder originating from body processes) and there has been pressure in some areas for the formulation of detergents to be changed to obviate the need for phosphates.

It is possible to produce phosphate-free detergents by using nitrilotriacetic acid salts instead. Unfortunately, it now appears that these compounds may have undesirable side-effects in that their presence in sewage prevents the precipitation of heavy metals as sludges in sewage treatment. The effluents would then contain more soluble heavy metals than hitherto which would constitute a potential toxic hazard. Bearing this in mind and remembering that 4 to 5 mg/l of phosphorus in sewage is due to bodily processes the insistence on conversion to phosphate-free detergents does not seem advisable and has now been stopped in the US. Effluent discharges are not, of course, the only sources of nutrients since in rural areas nutrients present in the run-off from farmland will probably exceed the amount present in effluent discharges. Some form of control of fertilizer application may be necessary in the future to reduce the rate of eutrophication. The addition of fertilizers in pellet form close to the roots instead of blanketing the area with fertilizer would seem to offer prospects of a more efficient use of nutrients with less escaping in run-off. As the catchment becomes urbanized the affect of agricultural uses of nutrients diminishes and attention must be turned to effluent discharges as the main culprits of accelerated eutrophication.

POLLUTION OF GROUNDWATER

Although groundwater supplies are normally effectively purified by the straining action of the rock as water percolates through it, soluble impurities are not readily removed. Thus, although a groundwater would normally be free from any suspended matter it may contain impurities in solution. Nitrogen compounds in agricultural run-off and effluents are responsible for increased nitrate levels in many groundwaters. The use of soakaways for the disposal of domestic and industrial effluents may imperil groundwater quality unless there is an impermeable stratum between the disposal area and the aquifer. Spray irrigation of industrial wastes onto land is a method of disposal which is outside existing pollution control legislation although it can be responsible for the creation of severe groundwater contamination problems. The infiltration of liquid from refuse tips is also a possible source of pollution and all of these potential hazards are increased if the rock is fissured since polluted water may then rapidly reach water abstraction areas. In country districts where water is obtained from shallow wells or springs and sewage is disposed of by soakaways or cesspits which are no longer watertight, the presence of synthetic detergent in the water (indicated by the formation of a head of foam on a glass of water drawn from the tap) may give the first warning of pollution of groundwater by sewage. Fortunately, the detergent, being soluble, travels through the ground more rapidly than the suspended matter, such as micro-organisms, which tend to be slowed by the filtering action of the rock.

Accidental pollution of aquifers can have serious long-term consequences to the quality of the water in them. The pollution of the water supply of Montebello in California with 2, 4-D weed killer gives a good example of the hazards of groundwater pollution. An unsatisfactory batch of 2, 4-D was discharged to the sewer, passed through the sewage works and then reached the river. Some distance downstream river water supplied underlying gravel beds from which the town's water was obtained. Very strong tastes and odours were produced and persisted for five years despite an initial dilution estimated to be 1 in 10 million.

DISCHARGES TO ESTUARIES AND THE SEA

Large estuaries and the sea itself have, until comparatively recently, been looked upon as a useful repository for almost any unwanted waste from sewage to radioactive compounds and nerve gases. Many large conurbations have developed on estuaries and the estuary has provided a convenient place for effluent disposal, particularly for industrial wastes which, if discharged inland, would require extensive treatment. As a result, many tidal rivers such as the Tees, Thames and Tyne (once salmon waters) have become grossly polluted.

When fresh water flows down a river and meets the saline seawater little or no mixing may occur and the fresh water tends to flow out of the estuary and over the denser seawater but, with its exit impeded, the river water and its pollution load may oscillate up and down the estuary for several days before finally escaping

Water Pollution

to the open sea. In these circumstances the trapped effluent can be a nuisance and is a potential health hazard on the banks particularly if these are popular playgrounds for children. Even in estuaries which are not stratified by density differences their large volume to inflow ratio means that the retention time of effluent discharged to the estuary can be quite prolonged. The slow transport of effluent to the sea is likely to result in considerable deposition in the estuary of solids which, when re-suspended by flood flows or storm conditions, can exert high oxygen demands. Such deposits may also necessitate dredging operations to maintain shipping channels although most dredging is needed to remove deposits brought in by the tide.

Discharge of significant volumes of wastewaters to estuaries is not, therefore, always a desirable means of disposal from the conservation point of view although it may well be economically attractive to communities within easy reach of a suitable estuary.

Direct disposal of effluents to the open sea would seem to be an attractive proposition and is practised by many coastal communities. Unfortunately, many old systems discharge via short outfall sewers extending only to low water mark or a little beyond. The discharge is thus often washed back onto the beach or, perhaps more often, onto the beach of an adjacent community. Because of the tidal drift, wind action and the effect of land mass configuration it is usually necessary to discharge sewage at a considerable distance from the coast if beach pollution is to be avoided. Tracer studies for proposed outfalls using floats and dyes released from a variety of places have often indicated the need for an outfall several kilometres long which makes the scheme costly and difficult to construct in exposed waters. Adequate dispersion facilities must be provided at the end of the

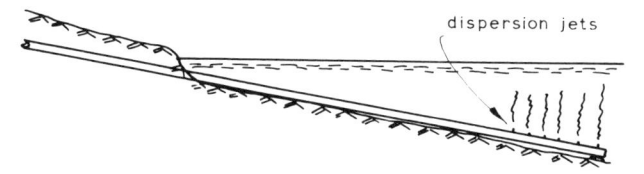

FIGURE 57. A sea outfall. Sewage is discharged through a number of disperser jets so placed as to ensure adequate dilution of the sewage before it reaches the surface

outfall (Fig. 57) to ensure rapid dilution and it is essential to remove, or destroy by maceration, the larger solids. This prevents the unsightly appearance of faeces, sanitary towels and contraceptives which are still the all too common indication of sewage pollution on many beaches today. The breaking up of large solids also exposes the bacteria to the unfavourable environment of seawater more rapidly and hastens their death. If investigations of flow patterns from a proposed outfall are not carried out under all weather and wind and tide conditions it is possible that certain combinations of circumstances may cause pollution of bathing beaches or shellfish beds in the vicinity. In this respect, it is interesting to note that pollution of estuaries has resulted in the abandonment of many shellfish beds. The shellfish filter large volumes of water during feeding and, thus, very efficiently collect such particles as bacteria, so that in polluted water they rapidly accumulate high populations of potentially harmful bacteria. The shellfish can then only be made fit for human consumption by storing them for some time in disinfected water in tanks before they are marketed.

There is little evidence of disease caused through bathing in polluted water but such waters and their beaches are certainly unattractive and a good deal of criticism has been recently directed to seaside resorts which have unsatisfactory sewage disposal facilities. Provided effluents are discharged in such a manner that adequate dilution takes place it is usually assumed that sea disposal is a satisfactory alternative to conventional sewage treatment. It might be, however, that in relatively shallow seas like the North Sea, and in land-locked seas, e.g., the Baltic, continued effluent discharges could give rise to ecological changes akin to the eutrophication of freshwaters. There has already been a sharp decline in DO levels in parts of the Baltic and some Scandinavian fjords show signs of eutrophication. Most of the new large electricity generating stations are built on the coast because their vast cooling water requirements could not be satisfied by inland waters and the return of large volumes of heated waters to the sea has resulted in marked changes in the biological populations near the stations. Large growths of organisms have been reported and in some cases the warm water in the vicinity of the discharge has attracted semi-tropical varieties of flora and fauna.

The ocean depths have long been used as a dumping ground for unwanted materials of war on the basis that there is little intermixing of the water in such regions and that in any case the dilution is very large. It does however seem that it is a rather dubious practice to dump hazardous materials in such a way that they are no longer under control and must eventually disperse throughout the oceans when the containers disintegrate.

CONTROL OF POLLUTION

As countries become more industrialized the need for pollution control becomes apparent and most countries now have pollution prevention authorities to control water quality in rivers and estuaries. Control of discharges to the sea is more difficult, since it requires international action but a start has been made in this direction with steps to combat oil pollution which has recently become a threat to many beaches in various parts of the world. The effect of oil discharges on bird life is particularly disastrous resulting in the death of many birds.

In England and Wales the quality and quantity of effluents discharged to surfacewaters and, in certain circumstances, into the ground are at present controlled by twenty-nine river authorities each being responsible for the catchment of a main river or group of rivers. In essence, there are two possible methods of controlling water quality.

(*a*) To decide on the desired quality of receiving water and, hence, determine the effluent standard required to maintain this quality in the river.

(*b*) To decide on a specified effluent standard based on the type of use to which the receiving water is put.

The first concept is more logical but leads to difficulties when additional effluent discharges are brought into use, since the standards applied to these would have to be more stringent than those applied to the existing discharges. In the second, the receiving water quality must inevitably deteriorate as additional effluents of the specified standard are added to the flow. In the UK, control of pollution is based on the recommendations of the Royal Commission on Sewage Disposal which proposed the adoption of effluent standards for BOD and SS in their eighth report (1912). It was then considered that a clean stream would normally have a BOD of 2 mg/l and that if the BOD rose above 4 mg/l the stream was close to becoming a nuisance. Since a fully treated sewage effluent usually had a SS of 30 mg/l and BOD of 20 mg/l (30:20 standard) it can easily be shown that such an effluent needs to be diluted eight times with clean river water to prevent the BOD exceeding 4 mg/l downstream of the effluent discharge. It should be noted that the Commission did not imply that at 4 mg/l the stream would not show evidence of pollution but simply that the stream would not be a public nuisance or a health hazard. The 30:20 standard has, until recently, been accepted as an almost universal effluent standard throughout the country although in many cases the dilution requirements have not been met so that considerable pollution has resulted. In situations with low dilution or where the receiving water is used for public water supply purposes downstream higher effluent standards may be required, e.g., 10:10 or even 5:5.

A recent survey of the quality of many rivers in the country has shown that most can assimilate a BOD of 4 mg/l without seriously affecting their use as raw-water sources or for fishing so that current pollution control practice is to adopt a 30:20 standard as the norm, unless it can be demonstrated that a higher effluent standard is necessary to preserve the quality of the river. As well as specifying BOD and SS values effluent standards usually stipulate the average and maximum rates of discharge and include limits for pH, temperature, etc. Constituents such as heavy metals, cyanides, etc., can also be controlled and limitations on ammonia and/or nitrate are sometimes imposed in cases where such compounds would be undesirable, e.g., public water supply abstraction downstream or discharges to lakes where eutrophication must be controlled. Table 13 shows typical river authority effluent standards.

Water Pollution

It is regrettable that, in spite of the control measures available, more than half of the sewage treatment plants in this country are not producing effluents which comply with their consent conditions. The main reason for these inferior effluents is overloading of many existing plants brought about by lack of finance for expansion and the delays which occur before new plant can be brought into operation. In some cases planning permission for new houses has been deferred until adequate treatment facilities are available but all too often there is a reluctance by river authorities to take action against other local authorities. As has often been said in local government affairs 'there are no votes in sewage treatment'; nevertheless, sewage treatment is an essential service which must be properly financed.

Industrial waste discharges tend to be even more unsatisfactory than those from local authority sewage works and, although prosecutions are more common, persuasion is the usual remedy and changes can take many years. Industrial wastes discharged directly to the sewer for treatment at local authority sewage works are often responsible for poor effluents from the works but, once the effluent has been accepted into a public sewer, the local authority then becomes responsible for any pollution which the waste causes. Because of this factor most local

Table 13. *Typical effluent standards for discharges to rivers* (concentrations in mg/l except where noted)

Characteristic	Maximum concentration
BOD	20
PV	15
SS	30
pH units	5–9
Temperature °C	25
Cyanide	0·1
Toxic metals	0·5
Phenol	0·5
Sulphide	1
Grease and oil	5

Table 14. *Typical effluent standards for discharges to sewers* (concentrations in mg/l except where noted)

Characteristic	Maximum concentration
pH units	6–11
Sulphide	10
Cyanide	10
Ammonia	100
Sulphate	1500
Grease and oil	500
Settleable solids	1000
Petroleum spirit	0
Chromium	50
Temperature °C	45

authorities will only accept industrial discharges into their sewers if they comply with certain conditions set so as to reduce the likelihood of the effluents interfering with the efficiency of the treatment works. Table 14 gives examples of standards required for industrial wastes to be accepted into public sewers. Limits are placed on constituents which might damage the sewer fabric or hinder treatment because of toxic properties.

Surveys of the state of rivers in England and Wales in 1958 and 1970 have produced the following information.

	per cent of total length	
	1958	1970
Unpolluted or recovering from pollution	72·9	76·2
Doubtful or needing improvement	14·3	14·7
Poor and urgently needing improvement	6·4	4·8
Grossly polluted	6·4	4·3

Although many rivers have improved in quality as the result of costly work on new or expanded treatment facilities it should be remembered that, because of the growing population and increasing demands for water, the proportion of sewage effluent in rivers is rising all the time. Control of effluent standards and the maintenance of river quality are thus becoming more and more important.

It would be unrealistic to assume that recent improvements in the quality of many rivers will continue without further intensive work in the pollution prevention field. Much of the recent improvement has been due to massive schemes for the construction of new treatment plants and the reconstruction of older works. Thus, the better effluents now being produced are due to what is almost a once and for all burst of construction. Indeed, the whole history of pollution control has a cyclic pattern—effluent load outstrips treatment capacity, then new treatment capacity is built so that, for a while, effluent standards improve but gradually deteriorate as the load again increases to exceed the available capacity.

9 Wastewater Treatment

From the previous chapter it will be clear that, in general, sewage and most industrial effluents require some form of treatment before they can be discharged to surfacewaters. The type of treatment necessary depends on the nature of the impurities to be removed and the degree of removal required.

COMPOSITION OF SEWAGE

In western communities each person produces about 0·06 kg BOD and 0·88 kg SS daily which, with a typical water consumption of about 200 l/person day, results in raw sewage of about 300 mg/l BOD and 400 mg/l SS. The total solids content of the sewage, usually about 1000 mg/l, depends in part on the dissolved solids content of the water supply to the area. About two-thirds of the total solids in sewage are organic in origin and the impurities are present as floating, suspended, colloidal and dissolved solids. Just as in water treatment, therefore, it is necessary to purify sewage by the use of a number of processes in combination to reduce the pollution load from 400:300 to the normal 30:20 effluent. The composition and volume of sewage depends on the type of sewerage system since with a combined system surfacewater will increase the flow and will also bring into the sewers considerable quantities of road grit, particularly in the winter. Thus, a treatment plant dealing with the discharge from a combined sewerage system will receive a maximum flow of 6 to 12 dwf whereas on a separate system the maximum flow would probably not exceed 3 dwf except on very small works.

SEWAGE TREATMENT

PRELIMINARY TREATMENT

The first operation in sewage treatment is to remove the grosser floating and suspended solids, paper, rags, etc. This may be achieved by passing the flow through a bar screen (Plate 27) with openings of 40 to 80 mm. On small works the screen is cleaned by hand but on larger installations mechanical screen-raking, operated by a time sequencer or by the increase in head

PLATE 27. A bar screen to remove paper and rags from sewage, debris being removed by a rake mechanism

loss across the screen as it clogs, is usually employed. The screenings are objectionable in nature containing a good deal of putrescible organic matter and their disposal may pose problems. Screenings are often buried but may be incinerated, alternatively a macerator may be used to reduce the screenings to a smaller size so that they can be returned to the flow for removal in the later stages of treatment. Instead of removing the material as

Wastewater Treatment

screenings a comminutor (Plate 28) can often be used to shred the solids to an acceptable size for removal later in the sedimentation tanks. The whole flow is passed through slots in a rotating drum so that large solids are caught by teeth in the slots and chopped by being brought into contact with a fixed cutter bar.

PLATE 28. A comminutor which may be used as an alternative to screening. The sewage flows in through the slotted drum which is slowly rotated. Teeth on the drum engage with slots on the cutter bar at the left and shred paper and rags trapped on the surface of the drum and in the slots

The grit present in the flow from combined sewerage systems would be troublesome in the main treatment plant since it would be likely to damage valve seats and pumps and so it is removed as part of the preliminary treatment process. The grit particles are much heavier than the other suspended matter and can be readily removed by a differential settlement technique. If the sewage is allowed to flow along a channel at a velocity of about 0·3 m/s the heavy grit will settle to the bottom but the organic solids will remain in suspension. The flow into a sewage works is continually varying over the range of say 0·4 to 9 dwf so that the grit removal channels must maintain the constant velocity over this flow range. This criterion can be achieved by using a channel of parabolic section controlled by a venturi flume which also serves to measure the flow (Plate 29). The settled grit is removed from the bottom of the channel by a bucket scraper or by suction and is usually washed to remove any adhering organic particles. An alternative method of grit removal is to use small sedimentation tanks with an overflow rate at maximum flow which removes grit particles only. At lower rates of flow some organic particles will settle out with the grit and must be washed back into the main flow. Such grit removal basins are more compact than grit channels but are not such an elegant solution to the hydraulic problem.

Full treatment is normally given only to a maximum of 3 dwf so that plants dealing with combined sewage utilize storm overflows (Plate 30) after grit removal to limit the flow to this value. Excess flows are diverted to stormwater sedimentation tanks (Plate 31) which are normally empty and only come into use when rainfall causes the flow reaching the sewage works to exceed 3 dwf. If the storm is not prolonged the storm tanks may not fill so that no discharge occurs. During long periods of rain the tanks fill and are then permitted to discharge to a suitable watercourse inevitably causing some pollution although the excess flow has at least been subjected to sedimentation to remove a portion of the suspended matter. When the inflow to the plant has fallen below 3 dwf the contents of the storm tanks are pumped back into the main flow to receive full treatment.

PLATE 29. Inlet works at a small sewage treatment plant. The flow passes through hand-raked screens and then along parabolic-section grit channels which are controlled by venturi flumes. Duplication of units enables repairs and maintenance to be carried out without interrupting the flow

PLATE 30. A double side-weir overflow. Flow in excess of 3 dwf spills over the weirs and is conveyed to the stormwater tanks

PRIMARY SEDIMENTATION

The larger suspended solids in sewage are removed by sedimentation in the same way as floc particles are removed in water treatment. Again, various types of unit are in use, circular or rectangular tanks with mechanical sludge scraping devices (Plate 32) for most works although hopper bottom tanks are popular for small works since the steep side slopes obviate the need for scrapers thus outweighing the extra construction cost.

PLATE 31. A stormwater settling tank. These tanks are normally kept empty and accept the flow separated by stormwater overflows. A light scraper assists movement of sludge to withdrawal hoppers

PLATE 32. A rectangular horizontal-flow primary sedimentation tank. The discharge weir is in the foreground

Conventional overflow rates at 3 dwf are about 30 m³/m²d and the tanks are usually 2 to 4 m deep. A certain amount of scum and grease accumulates on the surface and is prevented from escaping by scum boards in front of the take-off weirs. The scum is periodically removed.

Primary settlement of domestic sewage removes about 55 per cent of the SS in the raw sewage and about 35 per cent of the BOD, the remainder of the SS being non-settleable colloidal solids and the remaining BOD being mostly soluble. The settled solids collect as a sludge in the bottom of the tank and must be directed towards the sludge draw-off point by scraper blades moving slowly along the floor of the tank so that the solids are not resuspended. Primary sludge is removed at intervals, often once a day, usually at about 4 per cent solids (40 000 mg/l) and is an objectionable material with a heavy faecal odour.

After primary sedimentation the sewage will contain about 150 mg/l SS and 200 mg/l BOD which might be acceptable for discharge into the sea or into certain estuaries but is by no means acceptable for discharge into a river. It is thus necessary to give the settled sewage further treatment to remove much of the colloidal and soluble organic matter.

BIOLOGICAL TREATMENT

Self-purification of water utilizes micro-organisms to break down organic pollutants, subject to the availability of oxygen. The principle of biological treatment of sewage is to use the same aerobic oxidation reactions as occur in streams but to produce more rapid results by providing optimum environmental conditions in the treatment plant. Thus, the presence of ample food in the form of carbon compounds, excess oxygen and sufficient nutrients such as nitrogen and phosphorus ensure peak growth of micro-organisms so that large numbers of organisms are available in the system to metabolize the organic matter.

In the early days of sewage treatment, application of sewage to land in sewage farms provided a convenient means of disposal since biological action in the soil stabilized the organic content of the sewage. Land treatment is not, however, an ideal process because the soil may become sick after continued applications of sewage due to the accumulation of grease and perhaps toxic metals. The area of land required is large since the portion dosed with sewage must be rested after a period to allow it to recover. The likelihood of odour and other aerial nuisances from land treatment is high and these disadvantages have resulted in the virtual abandonment of land treatment for sewage disposal.

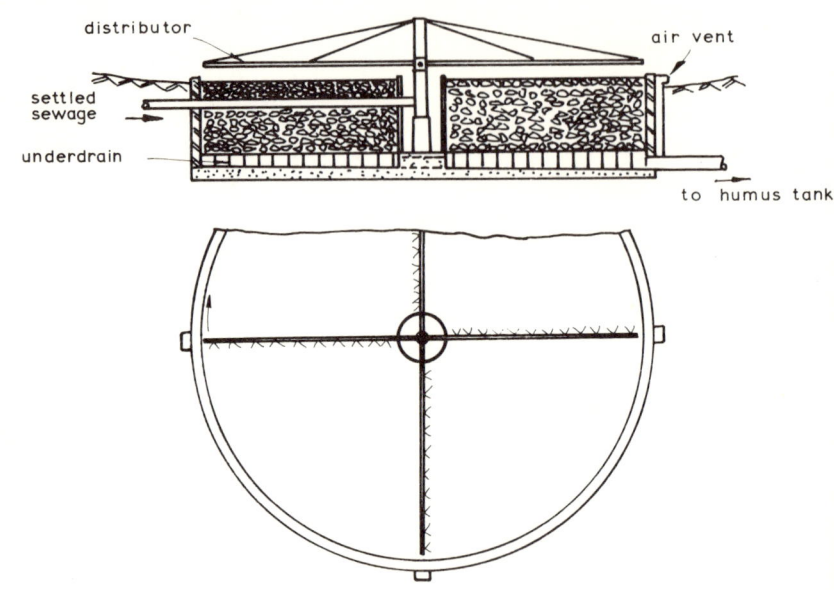

FIGURE 58. A bacteria bed

The original biological treatment process was the bacteria bed, otherwise known as the biological or percolating, filter although it is not a filter at all. Basically, the unit consists of a bed of stone or other angular medium, 50 to 100 mm size, circular or rectangular in plan and about 2 m deep (Fig. 58, Plates 33, 34). Settled sewage is intermittently sprinkled onto the surface of the bed and trickles down the space between the stones as air flows up from the ventilated underdrains. Micro-organisms attach themselves to the surface of the medium and produce a biological film which adsorbs the colloidal and soluble organic matter from the settled sewage. The organic matter is broken down biologically to give inert end products and allow the synthesis of new

Wastewater Treatment

PLATE 33. A small circular bacteria bed with reaction jet distributor. The concrete chamber in the foreground contains the dosing syphon

PLATE 34. Part of a large rectangular bacteria bed with rope-hauled distributors

micro-organisms. The end products are returned from the film into the liquid phase and are thus discharged from the bed. With normal strength sewage, the liquid reaching the bottom of the bed contains little organic matter in solution although a good deal of suspended matter is present mainly in the form of film which has been scoured off the medium due to the action of larvae and worms and the shearing effect of the passage of the liquid over it. The death of micro-organisms also leads to the film losing adhesion and if the film growth is excessive anaerobic conditions will occur at the boundary with the stones and adhesion will be reduced. Because of the suspended solids content filter effluent must be subjected to settlement in final sedimentation tanks referred to as humus tanks (Plate 35) to produce the desired 30:20 standard effluent.

PLATE 35. A circular humus tank. Filter effluent is conveyed to the central baffle chamber and settled effluent is discharged over the peripheral weir. Sludge is scraped to a central well by a blade suspended from the rotating bridge

If excessive organic loads are applied to a conventional bed the biological film increases in thickness to the point where the voids may become blocked and the bed becomes 'ponded'. Sewage cannot then percolate through the bed and the air flow is cut off so that purification stops and the bed becomes anaerobic.

High-rate filters using larger size stones or special plastics media which have high voids capacities are able to treat strong organic wastes without ponding at loading rates ten or twenty times the conventional loading of 0·1 kg BOD/m^{3}d. With high-rate filters, however, it is not normally possible to obtain such high percentage BOD removals as achieved by conventional units and they are thus unable to produce 30:20 standard effluents. High-rate units are useful for strong industrial wastes where a relatively small plant can give about 70 per cent BOD removal with the residual BOD being removed on a small conventional bed, the cost of the two units together being less than the cost of a single conventional bed to treat the whole load.

Sewage is fed onto the surface of the bed by some form of distributor mechanism, reaction-jet propelled on circular beds and rope-hauled arms on rectangular beds. During periods of low flow, e.g., at night, there may be insufficient flow to drive self-propelled distributors so that they would remain in one position giving a high load on the portion of the bed immediately below and allowing the rest of the bed to dry out with detrimental effects on the micro-organisms. To overcome this problem, small chambers containing syphons are inserted between the sedimentation tanks and the bacteria beds. These tanks hold sufficient sewage to operate the distributors for perhaps five minutes, at the end of which the syphon breaks and no further flow passes to the bed until the chamber has refilled. Provided the period between applications of sewage to the bed does not exceed about 30 minutes, no harm to the biological activity will occur. Bacteria beds are relatively lightly loaded and are thus able to absorb shock loads without too much difficulty. They also have the advantage of rarely needing external power and they require little maintenance. In these times of high land values the area required for bacteria beds and the high cost of the media together with the large head loss (about 2·5 m) through the beds restricts the number of sites at which they can be economically installed.

The other major biological treatment method is the activated-sludge process which is now adopted for most new works with contributing populations in excess of about 15 000 persons. The activated-sludge process uses the same basic reactions as occur in a bacteria bed but employs a high concentration of micro-organisms in the form of a flocculent suspension. The solids are kept in suspension by agitation with compressed air (Plates 36 and 37) or by mechanical aeration devices (Plates 38 and 39) which also ensure that ample dissolved oxygen, usually 1 to 2 mg/l, is present in the tank. Higher rates of oxygen transfer are possible with the activated-sludge process and thus a more heavily loaded plant can be built. The effluent from the aeration stage is low in dissolved organic matter but contains high concentrations of suspended solids, usually in the range 2000 to 8000 mg/l which must be removed in final sedimentation tanks. A portion of the sludge removed, which is largely composed of living organisms, is returned to the aeration stage where it can metabolize more organic matter (Fig. 59).

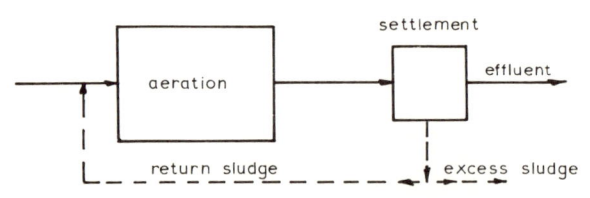

FIGURE 59. The activated-sludge process

Conventional activated-sludge units are loaded at about 0·6 kg BOD/m^{3}d and, since the depth of the aeration tanks is usually about 3 m, the area required is about an eighth of that required by bacteria beds to treat the same organic load. Head loss through the activated-sludge plant is small (less than 0·3 m) so that its use greatly eases the problems of site selection. Capital costs of activated sludge are lower than for bacteria beds because of the high cost of media. On the other hand, operating costs are higher for activated sludge and more skilled attention is necessary for efficient operation of the plant. As with bacteria beds, various modifications of the original activated-sludge process have been introduced to deal with strong wastes or to suit particular requirements.

In warm climates biological treatment is sometimes achieved in oxidation ponds which are shallow lagoons fed with raw or

Wastewater Treatment

PLATE 36. A diffused air activated-sludge plant

PLATE 37. One of the tanks shown in Plate 36 whilst under construction. The air supply enters the tank via the large vertical pipes on the left and is released through porous ceramic diffusors which are attached to the circular plates on the pipes in the bottom of the tank. The diffusors are more widely spaced towards the end of the tank because near the outlet the oxygen demand has been largely satisfied

PLATE 38. A rotating cone used to provide surface aeration as an alternative to diffused air. The turbulence created by the cone keeps the sludge in suspension as well as introducing air to satisfy the oxygen demand

PLATE 39. A brush-type surface aerator

settled sewage. The organic matter is broken down by bacteria, the end-products being utilized by algae in photosynthetic reactions which produce sufficient oxygen to keep the system aerobic. Under good conditions high-quality effluents can be produced from oxidation ponds but they are only suitable in locations where strong sunlight is available. They cannot be used satisfactorily in temperate climates.

FINAL SEDIMENTATION

Both main types of biological treatment process require sedimentation to remove suspended matter from the oxidized effluent. Tanks basically similar to those used for primary sedimentation are normally employed although at a higher loading of about 40 m^3/m^{2}d at 3 dwf. Because of the lighter and more homogeneous nature of secondary sludge simpler sludge scrapers are possible and scum removal is not normally necessary (Plates 40, 41 and 42).

PLATE 40. A circular final settling tank for the separation of activated sludge. The inset effluent weir provides more even flow patterns in the tank

TERTIARY TREATMENT

A conventional sewage treatment plant with primary sedimentation, biological oxidation and final sedimentation can reliably produce a 30:20 standard effluent but higher standard effluents, e.g., 10:10 cannot normally be achieved by such a plant. If a higher standard effluent is required a further, tertiary, stage of treatment is necessary. Tertiary treatment is based on the removal of some of the suspended solids which escape from final settling tanks and this removal also provides a somewhat smaller reduction in BOD (because only part of the BOD is associated with SS).

Micro-strainers or sand filters as used in water treatment are often employed to provide tertiary treatment for effluents from a conventional works and can usually produce effluents of about 10:10 standard. Other methods of tertiary treatment include; grass plot irrigation which gives good results but requires a considerable amount of land, pebble-bed clarifiers (Fig. 60) which serve to aid flocculation of the fine solids which would otherwise escape from the tank, and 'lagooning'. Tertiary treatment should only be used to improve a good 30:20 standard effluent and cannot be expected to turn a poor effluent from an overloaded works into a high-quality discharge. It should be noted that tertiary treatment does not remove any dissolved solids and can only remove a fraction of the micro-organisms present, so that in no sense can sewage effluent, even after tertiary treatment, be considered as drinking-water. Those who claim to be able to drink the effluent from a sewage works are either foolish or have very strong constitutions!

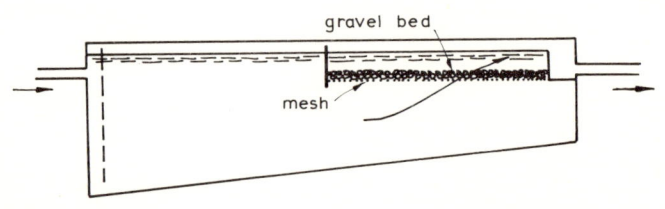

FIGURE 60. A pebble-bed clarifier. The shallow layer of pea gravel supported on mesh encourages flocculation and settlement of suspended solids which would otherwise escape from the tank

PLATE 41. Interior arrangement of the tank shown in Plate 40. The mixed liquor enters the inlet baffle via the vertical pipe and settled effluent escapes over the peripheral weirs. Sludge is withdrawn from the centre of the cone at the bottom of the tank

PLATE 42. Final effluent leaving the secondary sedimentation tank at a works producing a 30:20 standard discharge. The good clarity of the effluent is readily visible

SLUDGE TREATMENT AND DISPOSAL

The major problem in sewage treatment is the processing and ultimate disposal of the large volumes of sludge produced in sedimentation tanks. Indeed, up to half of the cost of sewage treatment may be accounted for by sludge processing and disposal.

Primary sludge is an objectionable material containing about four per cent solids largely in the form of putrescible organic matter which produces a heavy faecal odour. The biological oxidation phase of treatment produces about 0·5 kg of volatile solids for each kg of BOD destroyed in the process so that large volumes of sludge are collected in the secondary sedimentation tanks. This secondary sludge is low in solids content, usually about one per cent, and although mainly organic in nature, it is not so objectionable as primary sludge since the organic matter is present in a more stable form. Surplus activated sludge soon develops an objectionable odour since the unsatisfied oxygen demand results in the onset of anaerobic conditions. Sewage sludges contain highly compressible solids so that removal of the water surrounding them is often difficult. Rapid removal of the water results in compression of the solids which then present an almost impermeable barrier to the passage of any further water. Nevertheless, because of the large volumes of water associated with the solids it is important to remove as much of the water as possible to reduce the volume of material for disposal (Fig. 61).

In small works, the sludge is run onto drying beds (Plate 43) in a thin layer about 150 mm deep. Drying of the sludge then takes place, partly due to drainage through the sludge into the drains below the bed, the high organic content liquor being returned to the works for treatment, and partly by evaporation by wind and sun. Odour problems are likely if primary sludge is dried on beds although dosing with chemicals and the use of odour masking sprays can reduce the nuisance. In good conditions the sludge dries to a solids content of about 25 per cent in a few weeks and at this moisture content it has a soil-like consistency and can be lifted with a spade. With good weather a bed allowance of about 3 to 4 persons/m^2 is acceptable but in rainy

Wastewater Treatment

490 kg water

10 kg solids

2 per cent solids

30 kg water

25 per cent solids

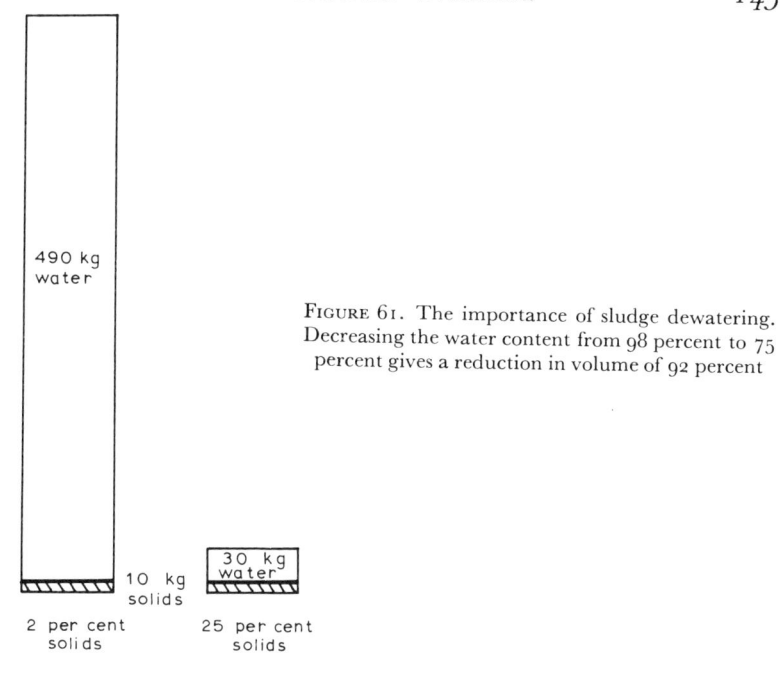

FIGURE 61. The importance of sludge dewatering. Decreasing the water content from 98 percent to 75 percent gives a reduction in volume of 92 percent

PLATE 43. Sludge drying beds at a small works. The upper bed has just received a charge of sludge but the lower bed has begun to dry out, cracks are forming which will speed up the drying process and the sludge will soon be ready for lifting

climates considerably more capacity is often necessary. The land required for sludge drying beds may be difficult to provide in large works and removal of large volumes of sludge must be mechanized adding further to the cost (Plate 44).

At the present time most sewage works digest their sludges anaerobically to produce a more stable end-product and to kill any pathogenic micro-organisms in the sludge. Digestion is carried out in closed tanks (Plate 45) heated to about 35°C by utilizing the waste heat from dual-fuel engines running normally on the methane gas produced in the process and which generate power for the works. Alternatively, the digesters may be heated by burning the methane in heat exchanger boilers. After a week or two, gas evolution falls off and the sludge is transferred to open secondary digestion tanks where separation of the solids occurs. The digested sludge is then drawn off as a black tarry-smelling suspension with about 4 per cent solids and the supernatant liquor containing a great deal of soluble organic matter (BOD of around 10 000 mg/l) is returned to the main treatment plant for stabilization in the aerobic oxidation units. This returned sludge liquor can account for as much as 10 per cent of the BOD load on the main plant. Digestion reduces the organic content of the sludge by about a third and produces a fine homogenized suspension which is sometimes rather more difficult to de-water than the original undigested sludge. Anaerobic digestion has a high capital cost and is susceptible to toxic matter in the sludge which may so reduce the rate of digestion, and hence the amount of gas produced, that external fuel has to be used to maintain the temperature.

In situations where de-watering on drying beds is uneconomic or undesirable mechanical de-watering is becoming increasingly popular, two main methods being available. Both have the advantages of occupying little space and being able to operate regardless of weather conditions so that large sludge storage facilities are not required. Pressure filtration is a batch process in which sludge, after chemical conditioning using doses of lime or aluminium chlorohydrate possibly aided by polyelectrolytes, is pumped slowly with increasing pressure into grooved plates supporting filter cloths (Plate 46). The cake formed at the surface

PLATE 44. A mechanical sludge lifter. This equipment, which is essential on large works, scrapes off the layer of dried sludge and discharges it via a system of conveyor belts to waiting lorries at the side of the beds

PLATE 45. A sludge digestion tank with floating gas holder cover. The tank is surrounded with earth to provide heat insulation and thus reduce the amount of gas required to maintain the contents at about 35°C

of the cloth acts as a filter for the remainder of the sludge and the filtrate is removed via the grooves in the plates. The pressing time is usually about four hours with pressures of up to 700 kN/m^2, the cake produced having a solids content of 25 to 50 per cent. Vacuum filtration is a continuous process in which a revolving segmented drum covered with filter cloth is partially submerged in a bath of conditioned sludge (Plate 47). A vacuum of about 90 kN/m^2 is applied to the submerged segments and a layer of sludge builds up on the cloth. As the sludge layer emerges from the bath air is drawn through it and this accelerates de-watering. The sludge cake, at about 25 per cent solids, is removed by a scraper blade which is assisted in its action by the application of a slight pressure to the segment as the segment approaches the blade.

Heat treatment of sludges is sometimes employed to stabilize the organic matter and also to de-water the sludge more readily. The Porteous process involves heating the sludge to about 190°C for 30 minutes at a pressure of about 1·5 MN/m^2, the sludge then being run into thickening tanks where the supernatant can be drawn off. Here again, as with all forms of de-watering, the liquid removed from the sludge is highly polluting and must be returned to the main works.

The disposal of sludge is also a problem at many works. Much sludge is dumped after de-watering in old quarries or similar sites but care must be taken to avoid polluting groundwater. In many areas, the number of sites suitable for sludge dumping is rapidly diminishing and, in any case, transport costs can make the whole operation expensive. Sewage sludge contains nitrogen and phosphorus but little potash (Table 15) so although it has some fertilizer value it is insufficient to qualify for a subsidy. Disposal of sludge to farmland either in the liquid state by spraying or as de-watered cake is practised by many authorities and is certainly desirable since nutrients are returned to the land. This means of disposal tends to be seasonal, and storage facilities must be available to hold production when access to the land is not possible. Care is also necessary when dealing with sludges from works dealing with industrial wastes since toxic materials such as heavy metals tend to appear in the sludge and could possibly be

Wastewater Treatment

PLATE 46. A modern filter process installation with 80-chamber mechanically operated presses

PLATE 47. A vacuum coilfilter unit which produces a continuous output of sludge cake from the drum

concentrated in food chains if applied to the land. Incineration of de-watered sludge is finding favour with many large authorities where tips are too far away and agricultural uses are not available. Very little residue remains after incineration and, provided efficient gas scrubbing removes objectionable matter from the stack gases, it offers the ultimate solution, at a cost, to the sludge problem. Several authorities dispose of sludge, usually in the liquid state, by dumping at sea from specially constructed vessels. On a large scale with full utilization of facilities, e.g., by sharing with other authorities, dumping at sea in deep water may well be the cheapest method of sludge disposal. Evidence from a dumping ground used for many years suggests that the sludge is rapidly dispersed on the ocean floor and presumably provides a source of food for the organisms in the sea. If the sludge contains solids from industrial wastes in significant proportions then the possible effects of such toxic materials as heavy metals should be considered before dumping is commenced.

Table 15. Composition of sewage sludges compared with a fertilizer

Constituent	Concentration as per cent of dry solids			
	Primary sludge	Activated sludge	Digested sludge	Fertilizer
Total nitrogen	4·5	6·0	2·0	6·0
Phosphorus as P_2O_5	2·0	2·5	1·5	10·0
Potash	0·5	0·7	0·5	6·0

COSTS OF SEWAGE TREATMENT

Figure 62 shows the flow diagram for a conventional sewage treatment plant, which will produce a 30:20 standard effluent at a cost of about 2p/m³, tertiary treatment to 10:10 standard adding about 10 per cent to the cost. The capital cost for a medium-sized conventional plant is in the region of £45/m³d.

Wastewater Treatment

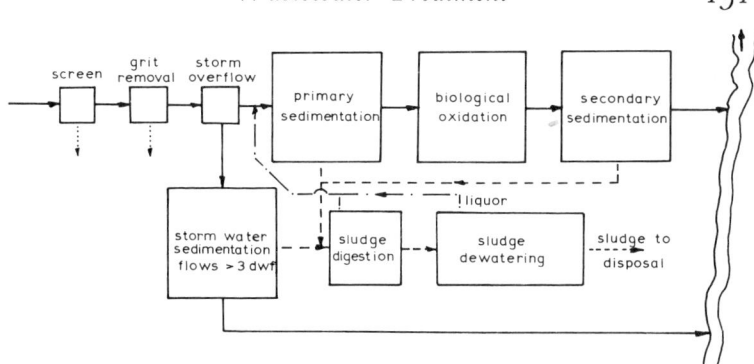

FIGURE 62. Flow diagram for a conventional sewage treatment plant

SMALL SEWAGE WORKS

For isolated houses and villages not connected to main drainage systems, small individual treatment plants are necessary. The septic tank is probably the commonest type and this consists of a settling chamber which collects suspended matter and allows it to decompose anaerobically. As a result, some of the organic matter is destroyed and the remainder goes into solution. Thus, the effluent from a septic tank is more-or-less free from suspended matter but has a high BOD and must be stabilized by application to a small-scale bacteria bed before being discharged into a stream. Regular de-sludging of septic tanks is essential since, otherwise, solids collect until the chamber is full and no further solids removal can take place.

In recent years, a development of the activated-sludge process in the shape of factory-built extended-aeration units (Plate 48) has become popular for small rural installations. These units work on the principle that by aerating the sewage for about 48 hours, instead of the six to eight hours used for conventional installations, some of the sludge produced by reduction of BOD is itself destroyed so that the volume of sludge from the plant is reduced. Because of the long aeration period the sludge is highly mineralized, i.e., much of the organic matter has been broken down and, thus, the sludge is relatively stable and does not give rise to odour problems even though it contains primary sludge (since this also has been subjected to aeration). The running costs of extended aeration plants are high compared with conventional plants but they are lower in capital cost for units serving populations of up to several thousands.

Many small sewage works give poor results because of lack of proper maintenance and the failure on the part of the owners to appreciate the way in which the plant operates. A good 'clean out' with disinfectant is not the best way to treat a biological treatment plant!

PLATE 48. A factory-built extended-aeration activated-sludge plant. This type of plant which is suitable for a small village is provided as a finished unit and only requires connection to the drainage system and an electricity supply. Duplicate air compressors are fitted to reduce the risk of breakdown

INDUSTRIAL WASTEWATER TREATMENT

Almost all industrial processes produce some type of liquid effluent, ranging from cooling water to the highly polluting wastes from milk processing. Local authorities are required to accept industrial wastes for treatment at municipal plants and

Wastewater Treatment

are empowered to make charges for acceptance and treatment of industrial effluents. As noted in Chapter 8 they can set standards on the waste before accepting it into the sewers. Thus, for example, a metal processing factory would not be permitted to discharge to the public sewer a waste containing high concentrations of cyanide or heavy metals which would be toxic to the biological units at the sewage treatment plant. The manufacturer must, therefore, remove these objectionable constituents before discharge to the sewer. If the discharge is direct to a watercourse the river authority would require him to remove objectionable matter to a much greater extent than would normally be required for discharge to the sewer. With some wastes with high BOD values the cost of treatment at a local authority works which uses conventional-rate biological treatment may be such that a cheaper alternative would be for the manufacturer to install a high-rate plant of his own to remove a large proportion of the BOD fairly cheaply and then pay the local authority only for removal of the small residual BOD.

Local authority charging schemes are usually built up from a number of components such as the following.

(*a*) Transport in the sewers.
(*b*) Primary sedimentation.
(*c*) Biological oxidation.
(*d*) Sludge treatment.

The charges for treatment being based on the treatability and strength of the waste as compared with domestic sewage.

Wastes containing organic matter are treated by biological means often by high-rate filtration on plastics media for strong wastes. For very strong wastes containing high organic solids, e.g., slaughterhouse discharges, high-rate anaerobic digestion followed by aerobic oxidation of the residual organic matter has been successful. Wastes containing inorganic constituents, e.g., metal finishing discharges etc., must be purified by other means. Hexavalent chromium, which is a highly toxic constituent of plating effluents, can be removed by precipitation following the addition of ferrous sulphate and lime. Alternatively, ion exchange treatment can be used to remove the chromium and this has the advantage that no sludge is produced and it is possible that chromium could be recovered from the waste stream after regeneration. The recovery of silver from photographic wastes is an established technique and serves the dual function of conserving the raw material and reducing effluent disposal problems.

Grease in textile effluents is often removed by precipitation after the addition of sulphuric acid and the recovered grease can be sold. Highly acidic or alkaline wastes must be neutralized prior to discharge and this is usually achieved automatically by dosing alkali or acid to produce a preset pH which is continuously monitored in the final discharge. In many cases, neutralization results in the production of large amounts of sludge so that sedimentation tanks and sludge handling facilities must be provided. Examples of wastes which require neutralization are kier liquors from textile and paper manufacture (highly alkaline) and metal finishing wastes (highly acidic). Table 16 shows the type of treatment necessary for a number of industrial effluents.

More intensive farming methods and growing concern at river pollution have focussed attention on waste discharges from agricultural operations. Factory-farming methods concentrate waste production in a small area and the traditional methods of disposal of manure on the land are often no longer feasible. The treatment of such strong wastes by conventional sewage treatment processes can, however, be very costly. Inexpensive oxidation plants using earth ditches rather than concrete tanks of which the Pasveer Ditch (Plate 49) is an example have been useful in some cases but there is little doubt that the cheapest form of disposal for the farmer is to the land and other methods should only be considered when land disposal is impossible. Seasonal wastes such as silage are again best disposed of to the land since their high strength would require expensive biological treatment plant even if the waste could be stored to spread the flow over the year (a biological treatment plant cannot be switched on and off at will and requires several weeks to develop its purification efficiency). The wastes arising from the processing of sugar beet are another problem since the season lasts only two or three months and produces large volumes of high BOD effluent. Most

Wastewater Treatment

Table 16. Treatment for various industrial wastewaters

Type of waste	Probable treatment
Coal washing	Settlement (flotation for dust)
Cyanide	alkaline chlorination
Dairy	aerobic biological oxidation
Distillery	aerobic biological oxidation (high-rate)
Farming	land treatment
Metal finishing	chemical precipitation or ion exchange
Slaughterhouse	anaerobic biological oxidation
Sugar beet	lagooning followed by aerobic oxidation

PLATE 49. A Pasveer ditch extended-aeration plant. Aeration is provided by rotating paddles which also serve to maintain sufficient velocity of flow in the ditch to keep the activated sludge solids in suspension

works now lagoon the effluent and feed it at a constant rate over the year to a conventional biological oxidation plant or, in some cases, to a municipal sewage works.

Since effluent treatment, which may be quite expensive, in no way helps a manufacturer to sell his products but only adds to costs it is hardly surprising that effluent treatment often receives low priority. It is up to the public at large and the pollution control authorities in particular to convince persistent offenders of the errors of their ways.

10 Water Reclamation

As outlined earlier, growing demands for water coupled with increased volumes of effluent discharged to surfacewaters means that, in the future, a water source is likely to be used much more extensively than is the case at the present time. The adoption of river regulating reservoirs rather than direct supply reservoirs results in the abstraction of water for public supply from the lower reaches of rivers which will contain significant amounts of treated wastewater from discharges upstream. Many lowland rivers already contain significant proportions of sewage effluent, e.g., the Lee which supplies part of London's water is half sewage effluent in the upper reaches and in dry weather the Thames is about one-third effluent at Teddington. Sewage effluent is being used today for a number of industrial process-water uses quite satisfactorily and, indeed, may be a cheaper source of water than alternatives such as from rivers or canals. The sea has, until recently, been considered as an infinite but unusable source of water because of its salinity. However, ships have for many years obtained fresh water by distillation of seawater and in arid parts of the world communities already obtain their drinking water from the sea or from saline groundwaters as a routine, albeit costly, measure.

RE-USE OF SURFACEWATERS

In the re-use situation pictured in Fig. 63 the means of removing impurities can be considered as taking place in three units: sewage treatment, self-purification in the river and the water treatment plant. One use of water domestically drastically alters its composition as shown in Table 17, the most troublesome changes being the increases in organic content and in total dissolved solids. Dissolved solids such as chlorides are not removed

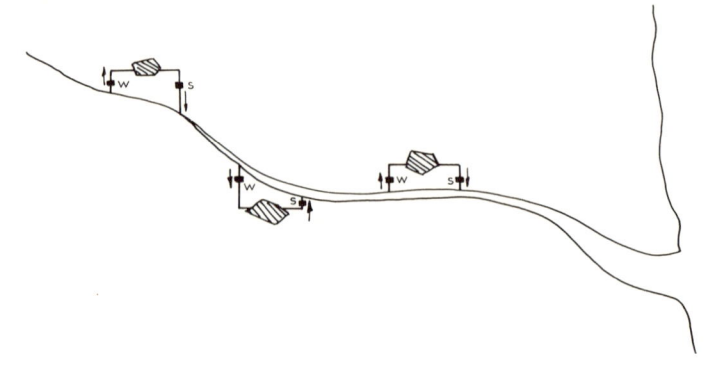

FIGURE 63. Reuse of water in a large river system. Each community abstracts water for treatment (w) and returns it as effluent after treatment (s)

Table 17. Quality degradation in one cycle, water to sewage effluent

Characteristic	Typical Increment: mg/l
COD	130
Organic nitrogen	5
Ammonia nitrogen	15
Nitrate nitrogen	5
Alkalinity	100
Total solids	250
Chloride	50
Phosphate	20
Sulphate	30

by any of the normal sewage or water treatment processes, nor are they removed during self-purification so the accumulation of dissolved solids is the limiting factor on the degree of re-use possible with conventional treatment methods. Chlorides may be detected by some consumers at about 250 mg/l, and most people will be able to detect a concentration of 500 mg/l so that such a level would be sufficient to render the water unpalatable. Since, for each water-sewage-water cycle, the chloride content increases by about 50 mg/l it will be clear that a definite limit is imposed on the number of cycles possible, depending upon the initial chloride content of the water.

Water Reclamation

Although most of the organic matter in wastewater is broken down biologically in the waste treatment plant or by self-purification in the receiving water there is always a residual of non-biodegradable organic matter which is of rather uncertain composition but which would tend to accumulate in a re-use situation. Some of this material might possibly be harmful to man but a more immediately obvious property is that it often causes tastes and odours in the water which are difficult to remove and it also increases the chlorine demand of the water. Nitrogen compounds accumulate in a re-use system either as ammonia or in the oxidized form as nitrates which can be harmful to young bottle-fed babies. Dissolved salts can be removed by a variety of methods which are dealt with later in this chapter in connection with the desalination of seawater. The organic impurities and nitrogen compounds found in re-cycled waters pose problems of their own which require specialized treatment methods.

REMOVAL OF DISSOLVED ORGANICS

It is not usually possible to identify all the residual organic compounds found in a polluted river since many are present only in trace concentrations. Their presence is, however, indicated by the chemical oxygen demand (COD) determination, by the carbon chloroform extract (CCE) procedure which uses the adsorptive power of activated carbon to remove organic matter from solution, or by determination of the total organic carbon (TOC) content. It has been proposed by the World Health Organization (WHO) that raw waters receiving conventional treatment should not have a COD exceeding 10 mg/l nor a CCE exceeding 0·5 mg/l. Since sewage effluents of 30:20 standard may contain more than 100 mg/l COD it is clear that a river receiving significant amounts of effluent will be unlikely to have a COD which would be acceptable under the proposed WHO criterion.

Removal of dissolved organic compounds can be achieved by the use of adsorption techniques. Adsorption is the phenomenon in which a solid surface in contact with a solution tends to accumulate a surface layer of solute molecules. Since adsorption is a surface action good adsorbents have a high surface-area:volume ratio and thus porous materials are usually good adsorbents. Activated carbon which is made by special treatment of hydrocarbon or carbohydrate compounds has a highly porous structure and has been used in powder form in the water industry for many years as described in Chapter 5. With waters regularly containing more than small amounts of organic matter the use of powdered carbon becomes uneconomic and granular carbon in beds similar to rapid gravity filters must be used. Such beds give efficient removal of almost all organic matter and produce a finished water with no taste or odour and little colour. The adsorptive capacity of the carbon eventually becomes exhausted and then no further removal occurs. The carbon must then be removed from the bed and regenerated by firing at 1000°C after which about 90 to 95 per cent of the adsorptive capacity is recovered. At least one water treatment plant in the US uses activated carbon beds with on-site regeneration facilities. Pre-treatment of the water by conventional coagulation, sedimentation and filtration prolongs the life of the carbon beds and reduces the operating cost. The cost of carbon bed treatment including, regeneration, is likely to be about $1p/m^3$, depending upon the quality of the feed water.

The foaming of detergents which beomes evident at concentrations of around 0·5 mg/l, can be utilized to remove organic matter from solution. By inducing foaming in a special tower with intense water aeration a large amount of foam results and this can be skimmed off the top of the tower. The foam contains most of the detergent present in the water as well as some of the organic compounds and its removal reduces the COD of the water by up to one-third.

REMOVAL OF NITROGEN COMPOUNDS

Sewage effluents may contain nitrogen compounds at concentrations of up to 40 mg/l, the form in which they occur depending upon the degree of treatment received. Many river authorities ask for nitrified sewage effluents in which most of the nitrogen is present as nitrate since, for this stage of oxidation to have been reached, almost all of the carbonaceous organic matter must have been stabilized. If ammonia is present in the discharge it will nitrify in the river and consume oxygen. From the drinking water point of view nitrogen in either form is undesirable and it is better to remove it from the system altogether.

Under anaerobic conditions, denitrification occurs and nitrate is reduced to nitrogen which is lost to the atmosphere. This reaction can be achieved by blending treated effluent with settled sewage in the ratio of about 3:1, under anaerobic conditions. The nitrate in the fully treated effluent is then reduced to nitrogen and the oxygen released can be utilized to oxidize the organic matter in the settled sewage. The final discharge is thus low in organic matter and shows a reduction in total nitrogen of about 35 per cent compared with conventionally treated effluent. Under alkaline pH conditions it is possible to remove ammonia by aeration in a stripping tower but relatively large volumes of air are needed for significant removals to be obtained.

REMOVAL OF PHOSPHORUS

Phosphorus compounds present in sewage at about 10 mg/l are partly due to natural body processes and partly to the phosphates which are incorporated in synthetic detergents. As we have seen the presence of phosphorus compounds in effluent discharges to lakes and slow-flowing rivers is undesirable because of their action in promoting algal growths. Phosphates are readily removed by coagulation with alum or lime although relatively high doses are necessary. Removal of greater than 95 per cent of phosphorus can be achieved with chemical doses of 200 to 400 mg/l. The high pH produced by lime treatment can also be utilized to give optimum conditions for ammonia removal by air stripping and thus reduce the overall cost of nutrient removal to some extent. As discussed earlier there is some doubt as to whether phosphate removal from effluents is a necessary operation in normal circumstances.

EXAMPLES OF WATER RECLAMATION

Two treatment plants which have recently been constructed for different reasons but with similar aims, namely to produce high-quality water from sewage effluent, are worthy of description.

(a) *Lake Tahoe Water Reclamation Plant.* Lake Tahoe in California is noted for the exceptional clarity of its water due to the barren nature of its catchment which has prevented the onset of eutrophication. Because of growing recreational activity in the area it was feared that waste discharges from a conventional treatment plant would impair the clarity of the lake by promoting algal growths. A water reclamation plant has thus been constructed to deal with the effluent from an 11 000 m^3/d capacity-activated sludge works. The treatment given includes coagulation and filtration followed by granular activated carbon beds and, finally, chlorination. Partial removal of ammonia is also practised by air stripping. The total cost of reclamation is about 3·5p/m^3 and, although the water is not used for domestic consumption, it would appear to meet the accepted standards for potable supplies. The reclaimed water is the sole source of supply for a recreational lake the discharge from which is used for irrigation.

(b) *Windhoek Reclamation Plant.* The further development of the town of Windhoek in S. W. Africa became dependent on the availability of additional water sources to augment the existing upland catchment and groundwater supplies. In the absence of satisfactory alternatives it was decided to reclaim water from the municipal sewage works effluent. The effluent after biological filtration is stabilized in oxidation ponds where algae take up the nutrients. The algae are then removed by coagulation and flotation and the effluent from this stage is treated in a conventional water treatment plant supplemented by foam fractionation and activated carbon adsorption beds. The reclaimed water, which costs about 4p/m^3 is blended in the ratio 1:2 with reservoir water to provide a supply of acceptable potable quality.

FRESH WATER FROM THE SEA

The use of seawater to meet future water demands is a solution almost invariably proposed by opponents of reservoir schemes. Whenever a new reservoir scheme is announced, objections are made asserting that flooding of the area would result in the loss of agricultural land, the dispossession of local inhabitants and damage to the scenery. The first two objections are certainly sometimes valid but it can be argued that the formation of a lake provides new recreational facilities and may often enhance the scenery if landscaping is carefully considered at the design stage. Nevertheless, the fact that fresh water can be obtained from the sea requires careful consideration of the techniques and costs of such treatment.

The sea apparently offers an almost limitless source of water ideally suited to an island community such as the UK. As we have seen however, the high dissolved solids content of about 35 000 mg/l—mainly sodium chloride—makes seawater quite undrinkable. In order to produce a potable supply from seawater it is thus necessary to reduce the dissolved solids to about 500 mg/l which is the desirable maximum for drinking water. In certain arid areas, notably the Middle East, brackish groundwaters containing about 10 000 mg/l dissolved solids are available but again need substantial removal of dissolved solids before they could be used as a potable supply. A number of methods of desalting saline waters are now available and are in use in many parts of the world.

DISTILLATION

Boiling of seawater in stills results in the evaporation of fresh water as steam which may be collected and condensed to provide distilled water of high purity. In fact, because of the almost complete absence of dissolved salts and gases, distilled water is unpalatable and it is usual to blend it with a small proportion of raw water and aerate the mixture before distribution. Shipboard supplies of drinking water have been obtained by distillation for many years and small portable distillation plants are often used to provide water for oil rigs and similar installations. In these circumstances other alternative sources of water, e.g., using part of the ship's cargo space to store fresh water or tankering in fresh water are costly so that distillation, although not cheap, is economic. When considering distillation or other desalination methods as alternatives to conventional water supply schemes in the UK costs are much more important. At the present time distillation costs are high compared with conventional water sources and, even if large dual-purpose nuclear power stations are used for power generation and desalination, costs are not yet competitive with conventional reservoir schemes.

In attempts to reduce the cost of distillation, new types of plant have been developed, the most efficient to date being the multi-stage flash process (Fig. 64, Plate 50). In this process, preheated seawater is fed into a series of chambers each of which is at a slightly lower pressure than the preceding one. Thus, water which has ceased boiling in one stage will again partially vaporize when passed into a lower-pressure stage. The vapour produced in each stage is condensed on tubes through which the seawater flows on the way to the first stage. In this manner as much heat as possible is recovered from the evaporation process thus, lowering fuel consumption and reducing operating costs. A large plant may have up to 40 stages with an operating temperature range of 90 to 50°C. At Guernsey in the Channel Islands a small 2250 m^3/d 12-stage plant is used to supplement the surfacewater resources of the island for the tomato crop where the additional value of the crop, due to better facilities for irrigation, outweighs the costs of de-salted water at about 12p/m^3. Many plants of similar size are in use in different parts of the world.

Design studies in 1965 for a large 625 MW nuclear power station which would produce 270 000 m^3/d of fresh water suggest a water cost of about 5p/m^3 if electricity is charged at 0·15p/kWh. If a higher electricity charge of 0·20p/kWh could be accepted the cost of water would fall to 3·5p/m^3. A plant of this type with a capacity of 190 000 m^3/d is at present under construction in Los Angeles. The relatively low costs given above only apply to very large stations and for smaller installations the costs increase

Water Reclamation

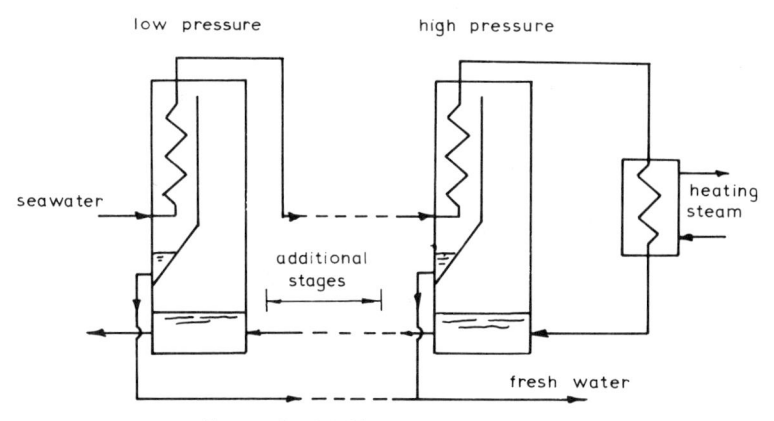

FIGURE 64. Multi-stage flash distillation

PLANT 50. A flash distillation plant for the desalination of seawater. The evaporators are in the centre with water storage tanks in the foreground

rapidly, e.g., a 14 000 m³/d dual-purpose plant would produce water at about 8p/m³.

Because of the high salt content of seawater and the presence of sulphates, corrosion and scale formation cause design and operational problems in distillation plants. As well as the freshwater stream from a distillation plant there is a waste stream of concentrated brine which requires suitable disposal arrangements. Care would have to be taken in estuarine sites that the waste brine did not eventually increase the salinity of the feed water.

Even a small distillation plant is a complex piece of equipment requiring skilled operation and maintenance so that attention has been directed towards other forms of distillation which might be more suitable for underdeveloped countries where skilled labour is in short supply. Solar energy can be harnessed to provide fresh water from seawater or brackish sources in a simple structure (Fig. 65). Unfortunately, the yield of such a device is low and the cost of glass panels is high. In addition, replacement of broken glass and the need to keep the glass clean means that considerable maintenance is required. Thus, the solar still does not seem to have as many advantages as might at first be imagined.

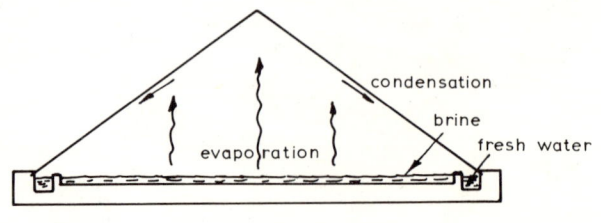

FIGURE 65. A solar still

ELECTRODIALYSIS

The only other desalination process in use on a full-scale basis at the present time is that of electrodialysis which is employed in over a hundred plants, the largest of which has a capacity of 10 000 m³/d. When a water containing dissolved salts is placed in an electrolytic cell and a potential applied some migration of ions occurs. The positive ions travel to the cathode and the negative ions travel to the anode. Because of leakage back from the electrodes the separation is not complete and it is not possible to produce de-salted water in the centre of the cell. When special ion-selective membranes are installed, however, desalination can be achieved. Using two types of membrane—one of which allows the passage of cations only and the other which is only permeable to anions—in a special cell (Fig. 66) water with less dissolved solids than the feed is obtained. Alternate chambers produce de-salted water and a stronger brine, which must be disposed of satisfactorily.

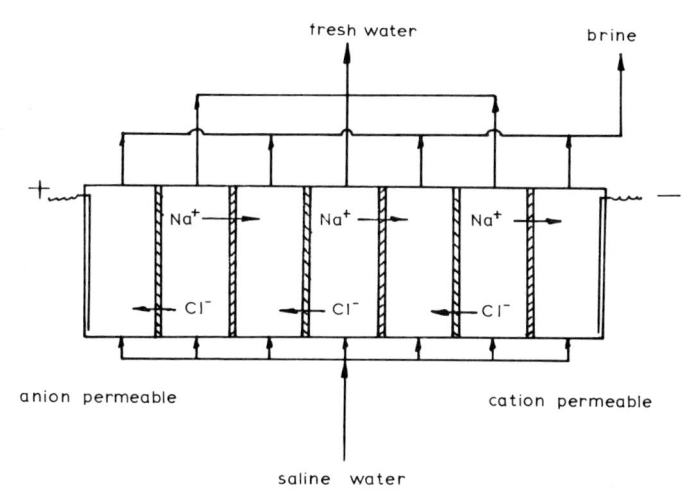

FIGURE 66. Principle of electrodialysis

The membranes are 0·5 to 1·5 mm apart to keep the electrical resistance, and hence the power consumption, as low as possible. In practice, electrodialysis is only economic for brackish waters with dissolved solids of up to 10 000 mg/l. The membranes have to be replaced after a period of use which may be several years if scale formation, corrosion and organic fouling can be prevented. Because of their modular construction electrodialysis plant costs are not so closely related to size as for distillation plants. For a brackish groundwater with 10 000 mg/l dissolved solids, reduction to 500 mg/l dissolved solids in a 90 000 m³/d electrodialysis plant would cost about 5p/m³. It has been estimated that to use electrodialysis to reduce the dissolved solids of seawater to 500 mg/l would cost about 60p/m³, a clearly uneconomic price.

REVERSE OSMOSIS

A process now under development based on the principle of osmosis appears to have considerable promise for de-salting seawater and also for purifying some types of industrial effluent. The phenomenon of osmosis is found with certain semi-permeable membranes, e.g., parchment, which will permit the passage of fresh water whilst preventing the passage of dissolved salts. Thus, if a semi-permeable membrane is used as a barrier between a salt solution and fresh water, the solvent (water) will pass through the membrane to equalize the salt concentration on either side. The transport of water occurs because of the osmotic pressure exerted by the dissolved salts. In reverse osmosis, a pressure in excess of the osmotic pressure is applied to the salt solution and water is transferred across the membrane giving a fresh water discharge and a brine waste stream. The process can be considered as one of hyperfiltration in which salt molecules are filtered out of the water (Fig. 67). With seawater, which has an osmotic pressure of about 2·5 MN/m², it is necessary to apply pressure of between 5 to 10 MN/m² to obtain a significant transfer of water. Because of these high pressures most of the initial work has been carried out on small-scale plants but commercial units are now on the market. The usual semi-permeable membrane is a form of a cellulose acetate which has little mechanical strength and, thus, must be physically supported by some form of backing which allows passage of the water but prevents rupture of the membrane under the high pressures used. A common form of support is to place the membrane inside a perforated copper tube or a porous fibreglass tube. Again, modular construction means that costs are not closely connected to the size of the plant. The membranes have a limited life and are capable of passing only a relatively low rate of flow

(0·5 to 2·5 m^3/m^{2}d) so that large areas of membrane are required. No full-scale costs are yet available for reverse osmosis but estimates, which may be a little optimistic, suggest that seawater could be de-salted for about 5p/m^3. There is some evidence to show that reverse osmosis can be used to remove metal ions from effluents and some membranes will remove organic compounds such as synthetic detergents.

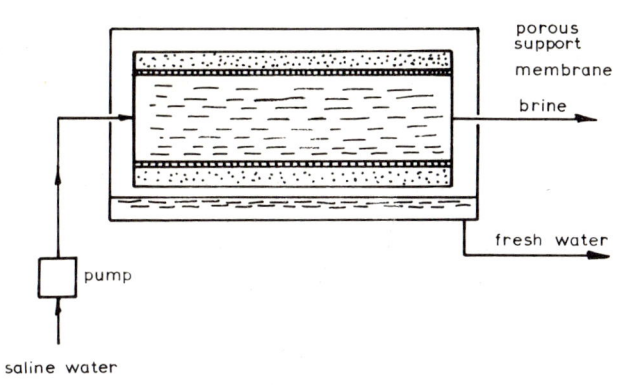

FIGURE 67. Principle of reverse osmosis

FREEZING

Desalination of seawater by freezing has several apparent advantages; thermodynamically it compares very favourably with distillation and at low temperatures there should be fewer problems with corrosion and scaling. Freezing processes depend on the fact that when brine is frozen the ice crystals which result are pure water although they are covered with a layer of salt. The crystals may be washed free of salt and then melted to give fresh water. The simplest method of freezing would be to use a heat exchanger through which a suitable refrigerant is passed. In such a system ice would build up around the walls of the heat exchanger, as it does in a domestic refrigerator, and would be difficult to remove. Thus, all work on freezing has been carried out by direct contact methods, i.e., by transferring heat directly from one fluid to another without a physical barrier. Two such processes are under development. *The Vacuum Freeze Process* (Fig. 68) removes heat from the salt solution by vaporizing a portion of the water under high vacuum. This water vapour is then condensed and becomes part of the product stream from the plant. Heat removed from the salt solution results in the formation of ice crystals which, when separated, washed and melted make up the remainder of the fresh water output. In the *Secondary Refrigerant Process* freezing of the salt water is achieved by vaporizing a secondary refrigerant mixed with the water, usually butane, which is then compressed and used to melt the ice after the washing stage. The butane vaporizes at nearly atmospheric pressure thus eliminating the need for special vacuum vessels. Contamination of the fresh water with butane is a possible disadvantage of this process. Costs for freezing are available only as predictions based on the performance of small-scale pilot plants and should be viewed with caution. It has been estimated that for a plant treating 4500 m^3/d of seawater the cost would be about $10p/m^3$. Here, again, disposal of the waste brine must not be forgotten. In spite of its theoretical advantages, freezing has not yet developed greatly as a desalination process, in particular there are many problems associated with the separation of the ice crystals from the brine. The Water Resources Board has recently abandoned work on a pilot scale freezing plant using butane which was to have been situated at Ipswich. Technical problems had resulted in considerable cost increases and it was concluded that the process was not yet sufficiently developed. Small scale experimental work on the process will, however, be continued.

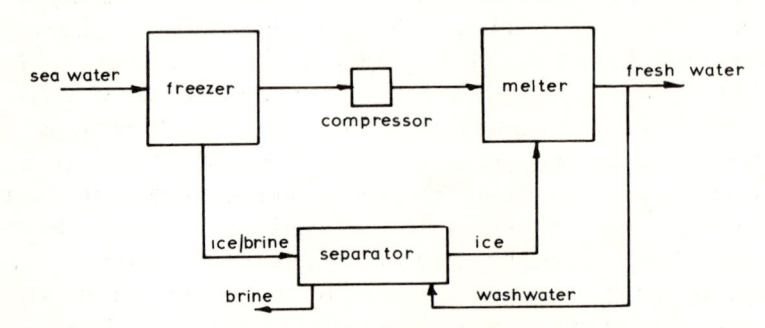

FIGURE 68. Flow diagram for vacuum freezing. Secondary refrigerant freezing is similar except that liquid butane is returned from the melter to the freezer for subsequent evaporation

11 *Economic and Legal Aspects*

So far, discussion has been limited to the technological aspects of water but consideration must also be given to the economic and legal situation which can place certain constraints on the use of technology.

ECONOMIC ASPECTS

It should be clear from what has been said in the proceeding chapters that the state of water science and technology is such that any water can be made fit to drink, at a cost. At the present time, the cost of water delivered to the consumer is about 2 to 3p/m^3 in the UK and at such a low cost it is perhaps inevitable that a good deal of water is wasted. In parts of the world where water is less plentiful, drinking-water may cost as much as 30 to 40p/m^3 and is used rather more circumspectly. It is often suggested that the rate of increase of water demand in this country could be reduced by introducing a more rational pricing scheme. Since the UK is one of the few countries which do not meter water for domestic consumption it is argued that the adoption of metering would result in a saving of water. The consumer would then pay for the amount of water used rather than a flat rate charge being levied regardless of the volume used. At existing charges for water, however, it has been estimated that the likely saving in water following the introduction of metering would not be of sufficient value to outweigh the cost of installing and reading the meters which would be of the order of £1·50 annually. As water costs rise it is likely that metering will become a more attractive proposition both to the water undertakings and to the consumer who might see it as a fairer way of pricing his supply. The whole concept of metering domestic water supplies depends on the thesis that water obeys the normal economic rules relating to the supply and demand of a commodity, i.e., the demand is related to the unit cost. Evidence from parts of the world where domestic metering is already practised does not however, always support this view, and there are examples of similar communities in the same areas in which more water is used per head in the community which has the higher unit charges.

The adoption of river-regulation reservoir schemes instead of direct-supply reservoirs clearly has benefits for the community at large as regards recreational and amenity use of the water. The water available for abstraction downstream does however, require more intensive and, thus, more expensive treatment than upland catchment water because of the other uses made of the water before it is abstracted. It is difficult to assess the monetary value of benefits such as angling or rambling alongside an attractive river so that a cost–benefit analysis for a river regulation scheme may be subject to a good deal of imprecision and doubt. There is perhaps some truth in the suggestions that sometimes the answer is decided first and the various costs and benefits are then adjusted to give the desired solution!

The subject of desalination as an alternative to conventional sources of supply is very much influenced by hard economic facts. Thus, distillation of seawater is likely to cost, at the most optimistic estimates, about $5p/m^3$ for treatment alone. To this treatment cost must be added the other supply costs due to additional link mains, service reservoirs and blending facilities necessary for any new source. These additional costs might well amount to a further $2p/m^3$ and in the near future, newly developed conventional sources are unlikely to have costs greater than about $3.5p/m^3$. At the present time therefore, and in the immediate future, desalination cannot compete with conventional sources of water on a base-load basis although the use of desalination as a supplementary supply may be attractive in certain situations. Holiday resorts have large seasonal influxes of visitors and, for a seaside community, seawater distillation could soon be an acceptable means of providing some at least of the additional water required during the holiday season.

Industrial consumers of water are likely to be more cost-conscious than domestic users and most of them already have metered supplies. Economy is not always practised, however,

Economic and Legal Aspects

and the sight of taps and hoses left running is still all too common in many industrial premises. It has also been the practice of some water undertakings to charge a lower unit rate to large consumers. This is, of course, in line with the procedures of other utilities like electricity and gas but it is not clear whether the aim of a water undertaking should be to persuade its consumers to use more water. Indeed, the electricity and gas boards must occasionally regret their advertising campaigns when growing demands result in the need to make voltage reductions or gas cuts. Water shortages and prohibitions on the use of hoses for gardens are often assumed by the general public to be an indication of inefficiency on the part of the water undertaking. In fact, such shortages are more often the result of the financial constraints placed on the development of water resources. The engineer can design facilities to cope with almost any demand situation but above a certain probability of occurrence it becomes uneconomic to cater for such a situation. The consumer must appreciate that if the frequency of water shortages is to be reduced from, say, three weeks every other year to one week every ten years much larger storage facilities will be required and, thus, water charges would be markedly raised. It is never possible to buy complete security from shortages because of the random nature of precipitation, a drought which statistically occurs once every 50 000 years may occur next year. A common fallacy is to believe that, because a structure has a life of 50 years, it will not have to cope with natural events like droughts or floods which have a return period of greater than 50 years. The probability of occurrence of, say, a 100 year event is the same in any year so that the event is just as likely to occur in the year that the structure is built as it is in the year in which the structure is written off.

Similar economic considerations govern the wastewater part of the hydrological cycle. Thus, it may be decided to give tertiary treatment to an effluent because it is discharged to a river which is utilized downstream for water supply purposes. Clearly, this decision produces an extra capital and operating cost at the wastewater treatment plant. At the present time this extra cost is normally paid by the drainage authority rather than the water undertaking. The fact that the higher quality effluent will probably improve the river and perhaps lead to better fishing does not appear in any accounting. There are several rivers in England whose dry weather flow is about half sewage effluent which therefore contributes a not inconsiderable benefit to other users. Indeed, cessation of effluent discharge to such rivers could have serious effects on the dry weather flow. Thus, although the river authority might press for higher standards it would be most reluctant to prohibit discharge of the effluent altogether as a pollution control measure. It could therefore be argued that the cost of effluent treatment should be at least partly offset by the value of the effluent to downstream users. It is perhaps worth noting that the water and wastewater industry in the UK has an annual expenditure of some £200 million and that treatment plants have a high ratio of capital invested to employees, much higher than many manufacturing industries.

LEGAL ASPECTS

Consideration of the legal aspects of water in all its forms is a lengthy and complicated matter and the following section is simply an attempt to outline some of the more important legislation found in England and Wales.

WATER SUPPLY

The first piece of legislation relating to water supply was the Waterworks Clauses Act of 1847. This required undertakings to provide, and keep in their mains, a supply of pure and wholesome water sufficient for the domestic use of all the inhabitants of the town and district within the limits of supply of the undertakers who were entitled to demand a supply and were willing to pay the authorized water rates. The supply was to be constantly provided at a pressure sufficient to serve the top storey of the highest house within the area unless other special arrangements were made. The third schedule to the Water Act of 1945 again stipulates a wholesome supply of water must be provided but, like many legislative instruments, it does not define its terms so

Economic and Legal Aspects

that the precise meaning of the word 'wholesome' as regards water supply is open to discussion. In this Act, the position regarding high buildings was clarified in that it was stated that undertakings are not required to deliver water at a height greater than that to which it would rise by gravity through the mains from the service reservoir. Thus, many modern tall buildings require booster pumps to raise water to storage tanks on the roof which are too high to be served by gravity flow.

Under the 1945 Act water undertakings are obliged to install and maintain hydrants for fire-fighting purposes as requested by the appropriate fire authority. Provisions for preventing waste and contamination of water are also contained in the Act which empowers undertakings to make bye-laws to regulate the construction and operation of all water fittings and to forbid the use of fittings which seem likely to cause waste or allow contamination of the water to take place. Water undertakings may also make by-laws to prevent pollution of surfacewaters and their catchments and also of groundwater.

As a result of the failure of the Dolgarrog Dam belonging to The North Wales Power Company and which caused 16 deaths in 1925, the Reservoirs (Safety Provisions) Act was passed in 1930. This Act requires all reservoirs holding 23 000 m^3 of water or more above ground to be inspected by a qualified engineer at intervals of not more than 10 years. The Act stipulates further that only a qualified engineer may be responsible for the design and construction of all reservoirs of this size.

As a result of the Water Resources Act of 1963, the Water Resources Board and the river authorities for England and Wales were set up. The river authorities are required to obtain data on the quality and quantity of water in their areas. They are empowered to construct works for the development of water resources and are able to finance this activity by licensing and charging for water abstraction in such a way that the water resources account is in balance. Water undertakings are, thus, no longer responsible for the development of resources which are now considered on a regional basis by the appropriate river authority.

SEWERAGE AND WASTEWATER TREATMENT

The Public Health Act of 1936 entitles the owner or occupier of any premises to have his drains connected to the public sewers so as to be able to discharge foul water, other than industrial effluent or prohibited liquids and surfacewater. Where separate sewers are provided foul sewage must not be discharged to the surfacewater sewer nor must surfacewater be discharged to the foul sewer except in special circumstances. The prohibited liquids under the Act are anything which may damage the sewer or interfere with the flow or treatment processes, liquids with a temperature exceeding 43°C, petroleum spirit and calcium carbide.

Industrial effluents were first covered by the Public Health (Drainage of Trade Premises) Act of 1937 which allowed such effluents to be discharged to the sewers with the consent of the local authority and provided for charges to be made for reception and treatment of the effluent. Discharges made prior to the date of the Act could continue to be made without the consent of the local authority provided the quantity discharged did not increase. This loophole in the Act, covering discharges with what was called a 'Prescriptive Right', was responsible for much difficulty at sewage treatment plants and a good deal of river pollution as a result. Effluents discharged under an agreement made prior to the Act were also exempt from its provisions. Laundries were exempt from the Act because it was felt that their discharges were only a substitute for home washing discharges although, today, with the large amounts of industrial washing carried out by laundries such an exemption would not seem so desirable.

The Public Health Act of 1961 empowered local authorities to charge for the reception and treatment of industrial effluents exempted under the 1937 Act, defined discharges from agricultural and horticultural premises as industrial effluents (much to the concern of farmers) and allowed the right of discharge for laundries to be withdrawn if the receiving sewer became overloaded as a result of the discharge.

The Public Health Act of 1936 states that it is the duty of every

Economic and Legal Aspects

local authority to provide such sewage treatment works as may be necessary for dealing effectively with the contents of its sewers.

PREVENTION OF POLLUTION

The major legislation governing the discharge of effluents to rivers is incorporated in the Rivers (Prevention of Pollution) Acts of 1951 and 1961 and the Water Resources Act of 1963. These Acts empower river authorities to give consent to the discharge of effluents into the rivers subject to compliance with quality standards and limitations on the total quantity and rate of discharge. The 1951 Act gave the then river boards the power to prevent the discharge of poisonous, noxious or polluting matter into rivers other than via public sewers and prevented the use of new or altered effluent discharges without consent. The 1961 Act brought under control discharges already in existence in 1951 and exempted from the earlier Act. The Water Resources Act extended pollution control measures to cover certain aspects of groundwater pollution in that discharges made to underground strata by means of well, borehole or pipe are subject to consent. An omission in the Act allows wastes to be discharged by simply spreading or spraying over the land, i.e., not by well, borehole or pipe, and has permitted some unsatisfactory situations to arise where groundwater has become polluted. There is now pressure from the river authorities and other interested bodies to alter the law to bring such disposal methods within the scope of the act. The disposal of radioactive wastes is controlled directly by the Department of the Environment under the Radioactive Substances Act of 1960 which lays down stringent conditions for the use and disposal of radioactive effluents and solid wastes.

The Clean Rivers (Estuaries and Tidal Waters) Act of 1960 gives river authorities power to control new and substantially altered discharges up to the seaward limit of an estuary. Under the 1951 and 1961 Rivers Acts, Tidal Waters Orders can be made to give river authorities full control of both new and existing discharges. The Sea Fisheries Committees around the coasts of England and Wales have no power to prohibit or control local authority discharges but do have by-laws to prohibit or control the deposition or discharge of solids or liquids detrimental to fish within the three-mile limit. There is no statutory control of dumping or discharge at sea beyond the three mile limit.

Returning to surfacewaters, it is of interest to note that under common law any riparian owner of land fronting onto a river is entitled to have the water in the river flow down to his land in the usual manner without sensible alteration in character or quality. There have been several successful common law injunctions gained against pollution by both local authority and industrial effluent discharges and it is no defence in such circumstances to plead that the offending effluent complies with a river authority standard.

It will be apparent from the foregoing summary of the legal position in this country that there are certain anomalies at the moment and new legislation will probably be introduced in the future to simplify the situation and close loopholes in the existing laws. The question of pollution of the sea from long outfalls and dumping is one which requires action on an international scale even though the hazards do not seem particularly great at the present time. Past experience shows that it would be much better to prevent serious pollution occurring rather than trying to clean up the mess after it has taken place.

12 Research and Future Developments

Because of the ever-increasing demands for water, conservation and proper development of the available resources is essential. In order to obtain the optimum use of water resources much more information is needed about their availability and character and also how natural and man-made activities affect these properties. Because of the low cost and ready availability of water in the past, relatively little research was undertaken into the overall concept of water resources management. Piecemeal utilization of water resources has been commonplace with each user considering his own requirement in isolation with little, if any, thought being given to its affect on the way in which future developments could be carried out. Today, the accent is on the multi-purpose development of water resources so that the greatest benefit can be achieved for the community as a whole. These widely-based schemes require thorough investigations in the planning stage to establish whether or not they are feasible propositions and to determine probable costs since, in most cases, economics will be the controlling factor.

CURRENT RESEARCH

The following examples illustrate typical methods of approach now being adopted in studies of water resources development.

THE TRENT ECONOMIC MODEL

The Trent with its catchment (Fig. 69) in the centre of England is a large river with great potential value for water supply purposes in the East Midlands since it flows from the wetter west to the drier east. The headwaters of the Trent are in the heavily industrialized West Midlands and most of the domestic and industrial wastewater from a population of two million is discharged to the Trent catchment. It is of interest to note that almost all of the water supply of the West Midlands is obtained, not from the Trent catchment, but from the Severn and Wye so that the Trent is a natural aqueduct conveying water from Wales to the east coast. Unfortunately for the East Midlands, the effluents discharged into the Trent and its tributary, the Tame, make it highly polluted and even after the quality is somewhat improved by the addition of the clean tributaries, the Dove and the Derwent, it is much more polluted at Nottingham than any river at present used for water supply purposes in the UK. Because of the future demands for water in the East Midlands, for which the volume of flow in the Trent would be a most useful acquisition to the water resources of the area, a complex analysis of the Trent system is now being carried out by the Water Resources Board, the Trent River Authority and other organizations.

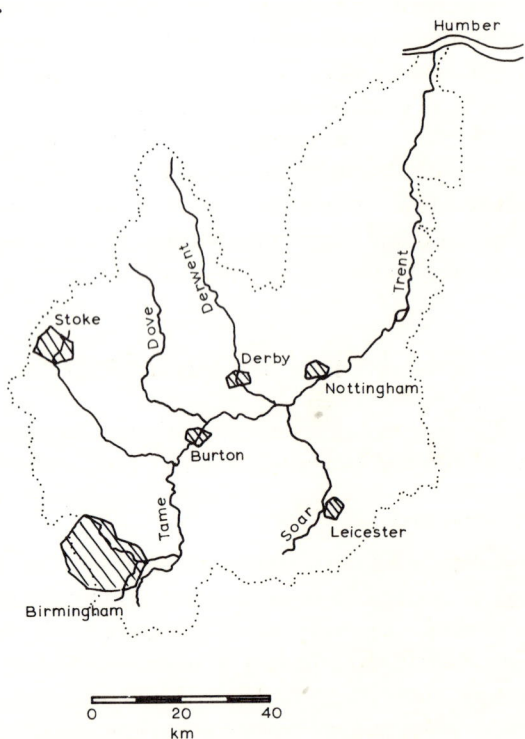

FIGURE 69. Catchment of the river Trent

Research and Future Developments

Future demands for water for domestic and industrial consumption and for recreational use are being assessed along with other resources in the catchment like groundwater reservoirs in the Bunter sandstone. Consideration is also being given to the idea of importing water from other catchments to the north and west. The methods available for treating Trent water to potable standards are under investigation in a pilot plant near Nottingham and work is also in hand to determine the most satisfactory way of treating the effluents presently discharged to the Trent and its tributaries. The possibility of treating the whole flow of the Tame in a river purification plant, as practised in certain heavily industrialized rivers in Europe, is also under investigation on a pilot scale. Costs and benefits of all the various factors are also being determined so that, by using a mathematical model to represent the behaviour of the Trent system, it will be possible to show the effect of all possible solutions and to determine the most economic solution. It remains to be seen whether the optimum solution on a technical and economic basis will prove politically acceptable. The work carried out on the Trent Economic Model is likely to be of great value when considering the use of other polluted river systems where extreme solutions could range from the acceptance of the river as an open sewer to the requirement that effluents discharged to the river must have extensive treatment including, perhaps, some measure of desalination.

AUGMENTATION OF FLOW IN THE THAMES

The Thames provides about two-thirds of London's water supply, and water for 24 other undertakings is drawn from the catchment of the Thames. The current demand is such that in a drought year the available resources would be insufficient and demand could only be met by reducing the flow over Teddington weir which is subject to statutory control. The Thames Conservancy has recently completed a three-year pilot study into the possibility of augmenting the summer flow of the Thames by abstracting water from the chalk and limestone in the western part of the catchment. Water from nine boreholes in the Lambourn Valley has been discharged into tributaries of the Thames whilst careful observations have been made of the hydrological behaviour of the area, the quality of the water and the ecology of the streams and their surroundings. The results have been encouraging and show that it would be possible to utilize numerous boreholes into the aquifers to provide water for flow augmentation in the Thames during dry weather. The aquifer would be replenished by natural percolation during the winter and would act as an underground impounding reservoir by preventing some of the winter run-off from being lost down the river. The Conservancy are now planning the adoption of such a scheme over a 10 year period. This scheme will substantially increase the dwf in the Thames at a cost about one-third that of conventional surface reservoirs and without the loss of any farmland, the inevitable result of the construction of surface reservoirs in the area.

RIVER DEE RESEARCH PROGRAMME

The Dee system, which incorporates two regulating reservoirs and provides water for seven undertakings and for canal use, has been subjected to an intensive investigation into methods of control. Much of the value of a river regulation scheme can be lost if the rules for discharge of water from the reservoirs are imprecise. For example, it is pointless to discharge water from a regulating reservoir to maintain flow at an abstraction point some distance downstream if, by the time the released water has reached the abstraction point, rainfall on the area has already increased the flow above the critical level. Clearly, such situations cannot be avoided altogether but more sophisticated control measures should be able to reduce their frequency of occurrence. Attention has therefore been directed towards the Dee catchment with the aim of producing better operating rules for regulation of the flow.

The need to obtain long-term rainfall and run-off data meant the development of methods for generating synthetic results based on the available records. By studying the statistical characteristics of rainfall or run-off data it is possible to produce a model for the data which will permit the generation of synthetic data which has the same characteristics as the original but can be made to extend over as many years as desired. In this way, it is possible to determine the magnitude of events which have a very long return period and which are unlikely to be included in the short-term records. This technique is finding increasing application in many water resources problems.

Improved rain gauge instrumentation has been installed in the Dee catchment and the use of radar for the prediction of rainfall is under investigation. Various systems of telemetry are being assessed to determine the most suitable methods for the transmission and processing of data from rain gauges, flow gauging stations, etc. Rapid and reliable handling of data is essential for efficient operation of a control system. The effect of artificial discharges of regulation water on fisheries, which are important in the Dee, has also been considered. As a result of the research programme it will be possible to construct a computer programme which will indicate how the discharge sluices at the regulating reservoirs should be operated to optimize the use of storage as well as preventing flooding downstream.

FUTURE DEVELOPMENTS

TRANSPORT OF WATER

As described earlier, one of the major problems in water resources is not an overall shortage of water but, rather, the situation where local demand exceeds the locally available resources. One solution is therefore to redistribute the available water by transporting excess flows from the wetter, less populated areas, to the drier, more densely populated, parts of the country. The river Trent as we have seen is a natural example of this principle which suffers from the disadvantage that the transport is not direct but via the water-sewage cycle, so that the quality of the water in the Trent suffers accordingly.

A man-made transport system recently completed diverts 110 000 m^3/d of water from the Ely Ouse in Norfolk via natural channels, newly-built tunnels and pumping stations into Abberton and Hanningfield reservoirs some 140 km away in Essex. The water is used to augment flow in the rivers Blackwater, Chelmer and Stour, thus alleviating water shortages in that heavily populated part of the country.

The source of the Thames in the Cotswolds is within a short distance of the river Severn and it would be a straightforward engineering task to pump water from the Severn into the Thames. Additional reservoirs would, however, have to be built in the Severn catchment to regulate the flow and permit increased abstractions. In the lower reaches of the Severn, from which the abstraction would be made, the quality is not particularly good. Thus, the groundwater scheme described earlier would seem to be a better solution for the present but recourse might have to be made to the Severn to satisfy requirements in the more distant future.

Proposals have been made for a Grand Contour Canal which would carry Scottish water down the Pennines to the heart of England as well as providing a waterway for goods traffic which would relieve congestion on the roads. While such a scheme may appear improbable it should be noted that artificial channels of similar size are already being constructed in California to bring water from the Sierras to water-hungry Los Angeles and southern California. It seems likely, therefore, that large-scale water transport schemes will become prominent in future water resources development.

BARRAGES

In recent years attention has been directed towards the possibility of constructing barrages across estuaries and bays to impound the freshwater inflow and, thus, provide a source of potable supply. A scheme of this type has already been completed in Hong Kong to ease the serious shortage of freshwater there. Using this technique, almost the whole flow of a river could be abstracted or impounded, since there would be no downstream use to be taken into consideration. Various sites have been subjected to feasibility studies by the Water Resources Board and their consultants, notably the estuary of the Dee, the Solway Firth, Morecambe Bay and the Wash. The problems

Research and Future Developments

entailed in the construction and operation of a barrage are enormous and at the present time it seems that the cost of such schemes in this country is likely to be considerably more than the cost of conventional upland reservoirs. If, however, the barrage itself can be used to provide communications, e.g., by road or railway links, some of the cost could be allocated to these aspects, thus lowering the water cost. The large freshwater lakes formed by barrages would be useful additions to recreational amenities and, possibly, this function could also absorb some of the costs.

The construction of a barrage is a large-scale civil engineering project but one for which there are precedents in Holland where large areas of land have been reclaimed from the sea in this way. The quality of water which would be obtained from a barrage scheme is, at the moment, rather problematical since most of the sites under consideration have large expanses of mud flats. It would presumably take some time for the salt to be leached out of these so that, initially, the stored water would tend to become saline. In addition, the reservoirs would be shallow and the storage of even moderately polluted waters under such conditions is likely to produce algal blooms which could be very troublesome.

Following investigations into a number of barrage sites The Water Resources Board has concluded that a barrage across the whole of the Wash is not feasible from the construction point of view. A barrage across the Solway Firth to supply the north would be more expensive than alternative conventional sources available in the area. Barrages across the Dee estuary or across Morecambe Bay might, however, be feasible to supply the additional needs of the Liverpool/Manchester area. A partial adoption of the barrage concept utilizing small pumped storage reservoirs reclaimed from the sea as polders would seem to offer possibilities as, again, almost the whole flow of a river could be impounded and the reservoirs would not occupy valuable farming land. It is envisaged that it might be possible to build three or four reservoirs in the Wash between the mouths of the Great Ouse and the Welland. Artificial walls, or bunds, about 10 m above sea-level would enclose the reservoirs which would be filled with fresh water by pumping from the rivers.

CONJUNCTIVE USE OF DESALINATION

Although the use of desalination plants to provide fresh water from the sea for base-load requirements will be uneconomic in the UK for some time a study by the Water Research Association suggests that the use of desalination in conjunction with conventional sources could be economic in some situations. The principle of this conjunctive use is to obtain higher yields from an existing catchment by overdrawing, i.e., by removing water from the reservoir at a rate higher than that which would prevent failure in, say, three consecutive dry years, and using desalination plant to make up the supply during dry weather when the conventional source would fail if abstraction were to continue at the higher level. Calculations for the Alwen reservoir in Wales show that, if the abstraction were increased by 20 per cent, additional water would only be required from the desalination plant for about 12 days a year on average. The cost of the additional 20 per cent yield would be only about half the cost of supplying all the extra yield from base-load desalination. This is because the reservoir costs are already partly written off and most of the cost for desalination is in fuel costs, the capital cost being relatively small compared with the construction of an additional reservoir to increase the supply.

POLLUTION PREVENTION

By the year 2000 the volume of effluent discharged in the UK will be about twice the present volume. This increase will be partly due to increased *per capita* water consumption so that the pollution load will not grow to quite the same extent as the volume of effluents. Nevertheless, it must be clear that by the end of this century surfacewaters will be receiving much larger pollution loads than at present and also that they will be drawn upon to provide water supplies to a much greater extent than they are today. If the quality of our rivers is to remain constant more stringent standards will have to be adopted, and if the quality of rivers is to be improved there will be a need for effluent standards of even greater stringency. Part of the present pollution load in rivers is due to the discharge of many effluents which do not comply with the 30:20 standard so that a first step must be to ensure that all discharges at least comply with this, relatively lenient, standard. To preserve river water quality against the onslaught of larger effluent discharges, tertiary treatment must become virtually universal, and the adoption of some form of desalination process, possibly reverse osmosis, will often be necessary to prevent the build-up of dissolved impurities. Inevitably, the costs of both water and effluent treatment will be increased.

ADMINISTRATION

At the present time responsibility for various stages of the water cycle rests with 160 water undertakings, 1200 sewerage authorities and 29 river authorities. The Government has recently announced plans for the reorganization of the administration of water resources which will involve the establishment of ten regional water authorities covering England and Wales. These new authorities are to replace the existing undertakings and authorities and will be responsible for water supply, prevention of pollution, water treatment, sewage treatment and disposal and the control of river flows. The regional water authorities, which are to come into operation in 1974 at the same time as local government is reorganized, will also be responsible for canals but are not to have responsibility for land drainage and fisheries. Revenue for the new authorities will come from charges for their services. The Water Resources Board is to be replaced by a National Water Council which will be of an advisory nature. The proposals should result in a more rational development of water resources and the piecemeal and often selfish developments of the past should not occur again under the new administrative system. Even today, the dividing line between a lowland river and an effluent which has received tertiary treatment is often indistinct. It is, thus, not illogical that a single water authority could combine the function of all the existing bodies. The practise of water science and technology could well be a fair description of the activities of such a multi-purpose water authority and it is clear that in the future the term 'water engineering' will have a much wider connotation than at present. The fact that water is one of the most important resources in the world must surely mean that water engineering should not be restricted to activities concerned with the supply and treatment of water for potable purposes which is, after all, only one part of the complex cycle.

Conversions to Imperial units

The units used throughout this book are those of the Système Internationale d'Unités (SI). Conversions to Imperial units are given below:

1 metre (m)	= 3·281 ft
1 m^2	= 1·094 yd^2
1 m^3	= 220 gal
1 litre (l)	= 0·22 gal
1 kilogramme (kg)	= 2·205 lb
1 m^3/s	= 19·01 × 10^6 gal/d
1 m^3/d	= 2190 gal/d
1 kN/m^2	= 0·145 lbf/in^2

For more detailed information see: *Metric Units with reference to water, sewage and related subjects*, HMSO, 1968.

Further reading

The following publications, all of which are likely to be readily available in the UK, provide more detailed information on various aspects of the subject matter of this book.

BOOKS

BARTLETT, R. E., *Public Health Engineering Design in Metric: Sewerage*, Elsevier Publishing Co., 1970.

BARTLETT, R. E., *Public Health Engineering Design in Metric: Wastewater Treatment*, Elsevier Publishing Co., 1971.

BOLTON, R. L., and KLEIN, L., *Sewage Treatment*, 2nd Ed., Butterworths, 1971.

ERICHSEN JONES, J. R., *Fish and River Pollution*, Butterworths, 1964.

FAIR, G. M., GEYER, J. C. and OKUN, D. A., *Waste Water Engineering*, Wiley & Sons Inc., New York, 1967.

HAWKES, H. A., *The Ecology of Waste-Water Treatment*, Pergamon Press, 1963.

HOLDEN, H. S., (Ed), *Water Treatment and Examination*, Churchill, 1970.

INSTITUTION OF WATER ENGINEERS, *Manual of British Water Engineering Practice*, 4th Edition, Heffer, 1969.

ISAAC, P. C. G., (Ed), *Waste Treatment*, Pergamon Press, 1960.

ISAAC, P. C. G., (Ed), *River Management*, Maclaren, 1967.

KLEIN, L., *River Pollution: 1. Chemical Analysis*, Butterworths, 1959.

KLEIN, L., *River Pollution: 2. Causes and Effects*, Butterworths, 1962.

KLEIN, L., *River Pollution: 3. Control*, Butterworths, 1966.

LUND, H. L., (Ed), *Industrial Pollution Control Handbook*, McGraw-Hill Book Co., New York, 1971.

MACAN, T. T. and WORTHINGTON, E. B. *Life in Lakes and Rivers*, Fontana, 1972.

McGAUHEY, P. H., *Engineering Management of Water Quality*, McGraw-Hill Book Co., New York, 1968.

McKINNEY, R. E., *Microbiology for Sanitary Engineers*, McGraw-Hill Book Co., New York, 1962.

SAWYER, C. M. and McCARTY, P. L., *Chemistry for Sanitary Engineers*, McGraw-Hill Book Co., New York, 1967.

SOUTHGATE, B. A., *Water: Pollution and Conservation*, Thunderbird, 1969.
TEBBUTT, T. H. Y., *Principles of Water Quality Control*, Pergamon Press, 1971.
TWORT, A. C., *A Textbook of Water Supply*, Edward Arnold, 1963.
WARREN, C. E., *Biology and Water Pollution Control*, W. B. Saunders Company, 1971.
WHITE, J. B., *The Design of Sewers and Sewage Treatment Works*, Edward Arnold, 1970.

GOVERNMENT PUBLICATIONS

DEPARTMENT OF THE ENVIRONMENT, *The Future Management of Water in England and Wales*, HMSO, 1971.
DEPARTMENT OF THE ENVIRONMENT, *Water Pollution Research*, (Annual Reports of the Director of the Water Pollution Research Laboratory), HMSO.
DEPARTMENT OF THE ENVIRONMENT, *Report of a River Pollution Survey of England and Wales 1970*, Vol. 1, HMSO, 1971.
MINISTRY OF HOUSING AND LOCAL GOVERNMENT, *Technical Problems of River Authorities and Sewage Disposal Authorities in Laying Down and Complying with Limits of Quality for Effluents more Restrictive than those of the Royal Commission*, HMSO, 1966.
MINISTRY OF HOUSING AND LOCAL GOVERNMENT, *Standards of Effluents to Rivers with Particular Reference to Industrial Effluents*, HMSO, 1968.
MINISTRY OF HOUSING AND LOCAL GOVERNMENT, *Taken for Granted*, HMSO, 1970.
WATER RESOURCES BOARD. *Annual Report*. HMSO.

JOURNALS AND PROCEEDINGS

INSTITUTION OF PUBLIC HEALTH ENGINEERS, *Journal*.
INSTITUTION OF WATER ENGINEERS, *Journal*.
INSTITUTE OF WATER POLLUTION CONTROL, *Water Pollution Control*.
SOCIETY FOR WATER TREATMENT AND EXAMINATION, *Water Treatment and Examination*.
Effluent and Water Treatment Journal, Thunderbird.
Water Pollution Abstracts, HMSO.
Water and Water Engineering, Colliery Guardian.
Water and Waste Treatment, D. R. Publications.

Index

Accidental discharges 47, 116
Acidity 6, 7
Activated carbon 81, 160, 162
Activated sludge 137–140, 144, 151, 152
Administration 187–188
Adsorption 70, 159–160
Aeration 138, 163
Aerobe 10
Aerobic oxidation 8, 10, 110, 113, 133, 153
Agricultural wastes 117, 118, 119, 154–155
Air scour 68
Air valve 90, 91
Algae 6, 11, 48, 57, 81, 111, 117, 141, 161, 162
Algicides 57
Alkalinity 6, 7, 28, 58
Aluminium chlorohydrate 146
Aluminium sulphate 58, 161
Ammonia 74, 75, 117, 124, 161, 162
Anaerobe 10
Anaerobic oxidation 8, 10, 46, 110, 136, 146, 153, 161
Angling 47
Anion exchange 80
Anthracite 72
Aqueduct 42
Aquifer 29, 48–51, 120
Arsenic 20
Asbestos cement pipes 87
Autotrophic organisms 10

Backwash 68, 69
Bacteria 11–14, 101, 110, 121, 133–141
Bacteria bed 134–137, 138, 151
Bacteriological analysis 11–14, 20
Baltic 122
Barrages 184–185
Bathing 111, 121
Bicarbonates 6, 77, 78
Bilham formula 32, 107
Biochemical oxygen demand (BOD) 8, 9, 28, 113–115, 117, 123, 124, 127, 133, 137, 138, 141, 144, 146, 151, 154
Biological characteristics of water 9, 14

Biological filter 134–137
Biological treatment 133–141, 146, 151–152, 154–155, 159
Blackwater, R. 184
Boiler feed water 7, 26, 80
Booster pump 175
Brackish water 163, 166, 167
Breakpoint chlorination 75
Bunter sandstone 48, 181

Calcium 77, 78, 79, 80, 83, 95
Carbon chloroform extract (CCE) 159
Carbon dioxide 6, 7, 8, 53, 78, 83
Carbonate 6, 77, 78, 83, 95
Cast iron pipes 86
Cathodic protection 95–96
Cation exchange 80
Chalk 48, 181
Charges for industrial effluents 23, 153
Chelmer, R. 184
Chemical characteristics of water 6, 14
Chemical conditioning of sludge 146
Chemical gauging 34–35
Chemical oxygen demand (COD) 9, 158, 159, 160
Chloride 9, 20, 157, 158
Chlorine 73–76, 77, 81, 82, 159, 162
Chlorine dioxide 81
Chlorophenols 4
Cholera 3
Chromium 153
Clean Rivers (Estuaries and Tidal Waters) Act 1960 177
Clywedog 43–45
Coagulation 57, 58, 59, 65, 66, 69, 160, 161, 162
Coal washing 155
Coliform bacteria 13
Collection of wastewater 99–108
Colour 5, 20, 58, 77
Combined residual 74–75
Combined sewer 99, 100, 127
Comminution 129
Compensation water 40
Concrete pipes 87, 102
Cone of depression 50
Conjunctive use 186

Index

Control of pollution 122–126
Cooling water 23–24, 111, 122, 152
Copper sulphate 57, 81
Corrosion 7, 83–84, 95–96, 101
Cost of treatment 54, 138, 150, 162, 164, 166, 168, 169, 170, 171–174, 186
Crimp and Bruges formula 102
Crustaceans 11
Current meter 33–34
Cyanide 9, 20, 124, 153, 155

Dairy wastes 155
Dams 41, 43, 44, 46
Dechlorination 74
Dee, R. 182, 183, 184, 185
Demineralized water 4, 26, 80
Density currents 62, 120
Derwent, R. 180
Detergent 110, 112–113, 118, 119, 120, 160, 161
Desalination 2, 157, 163–170, 186
Dewatering of sludge 65, 146–149
Dezincification 83
Digestion of sludge 146, 147
Dilution 121, 123
Direct supply reservoir 40–42, 43
Discrete particles 61
Disinfection 66, 73–77
Disposal of sludge 66, 148, 150
Dissolved oxygen 7, 50, 82, 110, 112, 113–115, 117, 122
Dissolved solids 5, 53, 111, 157, 159, 163, 170
Distillation 157, 163–166
Distillery wastes 155
Distribution of water 85–97
Domestic water supply 17–21
Dosing syphon 137
Dove, R. 180
Dredging 113, 120
Drying beds 144, 145, 147
Dry weather flow (dwf) 99, 127, 130, 141
Duplex brass 83

Ecological systems 9, 10, 109
Economic aspects 171–174
Effluent 115, 117, 119, 120, 123, 124, 125, 126, 127, 136, 142, 152, 153, 154, 155, 156, 157, 159, 160
Effluent standards 123–126, 186
Elan valley 41, 42
Electrical analogue 50, 51
Electrodialysis 166–168
Ely Ouse, R. 183
Epilimnion 46
Erie, L. 118

Escherichia coli 13, 21
Estuarine discharge 120–122, 133
Eutrophication 9, 117–119, 124, 162
Evaporation 14, 15, 38
Excess-lime softening 78
Extended aeration 151–152

Farm effluent 154–155
Ferrous sulphate 153
Fertilizer 119, 148, 150
Filtration 65–73, 160, 162
Filtration, biological 133–137
Final sedimentation 136, 141–143
Finance 124–125
Fireclay pipes 102
Fish 7, 44, 47, 111, 112, 116, 117, 124
Flocculation 57, 58, 59, 61, 62, 77
Floods 2, 37, 47, 99, 106, 183
Flow gauging 33–35
Flow-through curves 62
Fluoridation 82
Foam 112, 119, 160
Free residual 74–75
Freezing 169–170
Friction in pipes 86–87
Fungi 11

Gauging structure 35–36
Great Lakes 117
Great Ouse, R. 185
Grease 9
Greensand 78
Grit 127, 130, 131
Groundwater 5, 29, 54, 77, 81, 119, 148, 181
Guernsey 164

Hardness 7, 77–80
Hazen-Williams formula 86, 87
Head loss 70, 71, 72, 73, 86, 89, 138
Health hazard 111, 112, 120, 123
Heat treatment of sludge 148
Heavy metals 9, 20, 111, 118, 148, 153
Heterotrophic organisms 10
High-rate filters 137, 153
Horizontal flow tanks 65
Humus tank 136
Hydrogen sulphide 8, 27, 53, 74, 101, 104, 112
Hydrograph 36–37, 106, 107, 108
Hydrological cycle 14–16
Hydroxide 6, 58, 78
Hyperfiltration 168
Hypolimnion 46, 82

Impermeability factor 105
Incineration of sludge 150
Industrial wastes 6, 9, 20, 27–28, 111, 120, 125, 127, 152–156

Index

Industrial water supply 21–26
Infiltration 16, 104, 119
INSTAB 116
Intensity of rainfall 31–32, 106–108
Invert 101
Ion exchange 78–80, 83, 153
Ion selective membranes 167
Iron 58, 81, 153

Jar-test apparatus 58, 60
Joints in pipes 88, 102, 103

Lagoons 138, 141, 142
Lakes 44, 116–118
Lambourn valley 49, 181–182
Land treatment 133, 154
Langelier saturation index 83, 95
LD_{50} 116
Leakage 97
Lee, R. 157
Legal aspects 174–178
Lime 78, 83, 146, 153, 161
Lime softening 77
Lime-soda softening 78
Lining of pipes 95
Lowland water 5, 29, 39, 157

Maceration 121, 128
Magnesium 77, 78, 79, 80
Mains 85–89
Manhole 104
Mass-medication 82
Membranes 166, 168
Metabolism 10
Metal processing wastes 153, 155
Metals in water 7, 9, 20, 111, 118
Metering of water 19, 23, 171–172
Methaemoglobinaemia 118
Methane 8, 146
Micro-organisms 5, 9, 10–14, 57, 73, 109, 119, 133–141, 146
Microstraining 55, 142
Milk processing wastes 152, 155
Ministry of Health rainfall formulae 107
Mixed-media filtration 71, 73, 162
Moisture content of sludge 144, 145
Molecular attraction 70
Morecambe Bay 184, 185
Multi-purpose use of water 17, 179
Multi-stage flash distillation 164–166

National Water Council 187
Nitrate 8, 9, 48, 117, 124, 159
Nitrification 66, 69, 113, 170
Nitrifying bacteria 113
Nitrilotriacetic acid 118
Nitrogen 9, 20, 117, 118, 119, 133, 148, 159

North Sea 122
Nutrients 9, 48, 116–118, 133

Odour 122, 144
Oil 110, 116, 123
Oligotrophic waters 117
Ontario, L. 118
Operating costs 54, 150, 162, 164, 168, 169, 170
Organic compounds 4, 8, 27, 66, 74, 113, 133, 153, 157, 159, 160
Osmosis 168
Outfall sewer 120–121
Overflow rate 133, 141
Overturn 46
Oxidation pond 138, 141, 142, 162
Oxygen
 balance 110, 113–115
 deficit 114–115, 117, 122
 demand 8–9, 113, 120, 144, 161
 dissolved 5, 7, 50, 82, 110, 112, 113–115, 117, 122
 and fish 111, 112
 sag curve 115
 saturation 110, 111
Ozone 76–77

Pasveer ditch 154–155
Pathogens 10
Pebble bed clarifier 142
Percolating filter 134–137
Permanganate value (PV) 9
pH 6, 28, 58, 82, 83, 124, 154, 161
Phenol 4, 6
Phosphate 8, 9, 48, 78, 95, 117, 118
Phosphorus 9, 117, 118, 133, 148, 161
Photographic wastes 154
Photosynthesis 10, 115, 141
Physical characteristics of water 5, 14
Pipe
 friction 86–87
 joints 88, 102, 103
 linings 95
 materials 86–88, 102–103
Plastics media 137, 153
Plastics pipes 87
Plate counts 12
Plumbosolvency 83
Pollution 1, 3, 9, 21, 26, 47, 109–126, 186–187
Polyelectrolytes 59, 146
Ponding of filters 136
Power generation 146, 164, 166
Precipitation softening 77–78, 83
Prescriptive right 176
Pressure filtration of sludge 146, 149
Pressure zone 89, 96
Prevention of waste 19, 97
Primary sedimentation 131–133

Index

Protozoa 11
Public Health Acts 1936 & 1961 176
Public Health (Drainage of Trade Premises) Act 1937 176
Pumping 89, 91, 92, 93, 102, 103
Pyrolusite 82

Quality measurement 38–39

Radial flow tanks 65
Radioactive Substances Act 1960 177
Rainfall 1, 14, 16, 30–32, 99, 104, 182
Raingauging 30–36
Rapid
 gravity filter 67–69, 142
 mixing 58
 pressure filter 69, 70
Rats 104
Reaeration 114–115
Recharge 148
Reclamation of water 157–170
Recreational use of water 44, 46, 111
Refuse tips 119
Regional water authorities 185
Research 179–183
Reservoirs
 multi-purpose 42–45, 157, 172, 182
 single-purpose 40–42
 storage 47
 stratification 44, 46
 yield 40
Reservoirs (Safety Provisions) Act 1930 175
Reverse osmosis 168–169
Re-use of water 157–162
Riparian owner 178
River authorities 123, 175
Rivers 36, 39, 42–45, 47–48, 54, 157, 172, 182
Rivers (Prevention of Pollution) Acts 1951 & 1961 177
Rotifers 11
Royal Commission on Sewage Disposal 124
Run-off 14, 16, 29, 32–37, 47, 99, 100, 102, 104–108, 109, 182

Salinity 1, 2, 157, 163
Scour valve 90
Screening 55, 128–129
Sea discharge 120–123, 133, 177–178
Seawater 48, 52, 121, 157, 163–170
Secondary refrigerant freezing 170
Sedimentation 60–65, 70, 77, 129, 130, 131–133, 141–143, 144, 160
Self cleansing velocity 101
Self-purification 110, 113–115, 133, 158

Semi-permeable membrane 168
Separate sewers 99, 110
Septic tank 151
Service reservoir 96–97, 175
Settling velocity 57
Severn, R. 43, 44, 45, 180, 184
Sewage 3, 27, 28, 99, 109, 111, 127
 effluent 5, 48, 52, 115, 117, 119, 120, 123, 126
 farm 133
 fungus 112, 113
 treatment 128–156
Sewerage 99, 101–108, 127
Shellfish 121
Silver 154
Slaughterhouse wastes 153, 155
Slow sand filter 66
Sludge
 dewatering 65, 146–149
 digestion 146, 147
 disposal 65, 78, 148, 150
 heat treatment 148
 incineration 150
 pressing 147–148
 vacuum filtration 148
Small sewage works 151–152
Soakaway 119
Sodium carbonate 78, 83
Sodium hypochlorite 73
Soffit 101
Softening 77–80
Solar still 166
Solids
 colloidal 53
 dissolved 5, 53, 111, 157, 159, 163, 170
 floating 53
 suspended (SS) 28, 53, 113, 117, 123, 124, 127, 133, 142
 volatile 5, 144
Solway Firth 184, 185
Sources of water 51
Springs 48, 119
Spun iron pipes 87
Statistical analysis 29, 37, 182, 183
Steel pipes 87
Sterilization 73, 77
Stoke's Law 61
Stormwater
 sedimentation 130, 132
 sewers 104–106
 overflows 99, 100, 130, 131
Stour, R. 184
Stratification 46–47, 82, 120
Sugar beet wastes 154, 155
Sulphate 8
Sulphur dioxide 74
Superchlorination 75
Surface overflow rate 61

Index

Surfacewater run-off 99, 100, 105–108
Surge prevention 91–92
Suspended solids (SS) see solids suspended

Tahoe, L. 162
Tame, R. 112, 180, 181
Taste and odour 5, 6, 7, 9, 48, 81, 119
Tees, R. 120
Teeth and fluoridation 82–83
Temperature 6, 110, 122, 124
Tertiary treatment 142
Thames, R. 3, 112, 120, 157, 181–182, 184
Thames Conservancy 49, 181
Thermocline 46
Thiourea 113
Threshold concentration 115
Tidal drift 121
Time
  area diagram 107
  of concentration 106
  of entry 106
  of flow 106
TL_m 116
Total organic carbon (TOC) 159
Toxic compounds 20, 47, 111, 116–117, 125, 146, 148, 153
Tracer studies 121
Transpiration 16, 38
Transport of water 183–184
Treatability of industrial wastes 153
Trent, R. 179, 180, 181, 183
Trent Economic Model 179–181
Trent River Authority 180
Turbidity 5, 47, 59, 65, 66
Tyne, R. 112, 120
Typhoid 3, 21

Ultra-violet radiation 77
Upflow filter 71
Upland water 5, 29, 39, 54, 77

Vacuum filtration of sludge 148, 149
Vacuum freezing 170
Valves 90–91
Vertical flow tanks 63

Vibrio cholerae 3, 11
Viruses 11
Volatile solids 144

Wash 184, 185
Wastewater 27–28, 99–108
Water
  biological characteristics 9, 14
  chemical characteristics 6, 14
  collection 29–52
  conservation 1, 16
  cost 2, 54, 171–174
  demand 1, 2, 16, 18, 19, 21, 25, 29, 171–173, 181
  distribution 85–97
  domestic 17–21
  industrial 21–26
  mains 85–89
  physical characteristics 5, 14
  pipes 86–87
  pollution 109–126
  quality 38–39, 53–54
  reclamation 157–170
  sources 39–51
  towers 96–98
  use 17–28
  waste of 19, 23, 97, 173, 175
  wholesomeness of 19
Water Act 1945 174
Water Pollution Research Laboratory 116
Water Research Association 186
Water Resources Act 1963 175, 177
Water Resources Board 170, 175, 180, 184, 185, 187
Waterborne disease 2, 3, 9, 121
Waterworks Clauses Act 1847 174
Weed killer 120
Weil's disease 104
Welland, R. 185
Wells 3, 48, 119
Windermere, L. 118
Windhoek 162
World Health Organization 21, 22, 159
Wye, R. 180

Zeolite 78, 79